Monograph Series Volume 3

The Work of Mathematics Teacher Educators

Continuing the Conversation

Edited by

Kathleen Lynch-Davis
Appalachian State University

Robin L. Rider
East Carolina University

Monograph Series Editor

Denisse R. Thompson
University of South Florida

Association of
Mathematics Teacher Educators

Published by the Association of Mathematics Teacher Educators, San Diego State University, c/o Center for Research in Mathematics and Science Education, 6475 Alvarado Road, Suite 206 San Diego, CA 92129

www.amte.net

Library of Congress Cataloging-in-Publication Data

The work of mathematics teacher educators: continuing the conversation / edited by Kathleen Lynch-Davis, Robin L. Rider.
p. cm. – (Monograph series : v. 3)
Includes bibliographic references.
ISBN 1-932793-04-6
1. Mathematics–Study and teaching–United States. 2. Mathematics teachers–Training of–United States.
I. Lynch-Davis, Kathleen, 1972-. II. Rider, Robin L., 1963-. III. Series: Monograph series (Association of Mathematics Teacher Educators); v. 3.

Library of Congress Catalog Control Number: 2006032073
International Standard Book Number: 1-932793-04-6

Printed in the United States of America

The publications of the Association of Mathematics Teacher Educators present a variety of viewpoints. The views expressed or implied in this publication, unless otherwise noted, should not be interpreted as official positions of the Association.

Contents

Foreword
Sid Rachlin, East Carolina University v

Introduction

1. **Continuing the Conversation on Mathematics Teacher Education** 1

   Robin L. Rider, East Carolina University
   Kathleen Lynch-Davis, Appalachian State University

Teaching Preservice Teachers

2. **Using K-12 Mathematics Curriculum Materials in Teacher Education: Rationale, Strategies, and Preservice Teachers' Experiences** 11

   Gwendolyn M. Lloyd, Virginia Polytechnic Institute and State University

3. **A Rich Problem and Its Potential for Developing Mathematical Knowledge for Teaching** 29

   Judith Flowers, University of Michigan-Dearborn
   Rheta N. Rubenstein, University of Michigan-Dearborn

4. **Creating and Implementing a Capstone Course for Future Secondary Mathematics Teachers** 45

   Melissa Loe, University of St. Thomas
   Lisa Rezac, University of St. Thomas

5. **Renewing the Conversation about Gender Equity in Teacher Education** 63

   Lauriann Kress, Columbia University
   M. Lynn Breyfogle, Bucknell University

Teaching Inservice Teachers

6. Linking Teachers Online: A Structured Approach to Computer-Mediated Mentoring for Beginning Mathematics Teachers 79

Jennifer L. Luebeck, Montana State University - Bozeman

7. Teacher Learning in Mathematics Teacher Groups: One Math Problem at a Time 97

Sandra Crespo, Michigan State University
Helen Featherstone, Michigan State University

8. Blending Elements of Lesson Study with Case Analysis and Discussion: A Promising Professional Development Synergy 117

Edward A. Silver, University of Michigan
Valerie Mills, Oakland (MI) Schools
Alison Castro, University of Illinois-Chicago
Hala Ghousseini, University of Michigan

Teaching Teacher Educators

9. Transition to Teacher Educator: A Collaborative Effort 133

Laura R. Van Zoest, Western Michigan University
Diane L. Moore, Western Michigan University
Shari L. Stockero, Michigan Technological University

10. Designing Learning Opportunities for Mathematics Teacher Developers 149

Paola Sztajn, University of Georgia
Deborah Loewenberg Ball, University of Michigan
Teresa A. McMahon, University of Michigan

Rachlin, S.
AMTE Monograph 3
The Work of Mathematics Teacher Educators
©2006, pp. v-vii

Foreword

As President of the Association of Mathematics Teacher Educators, I am thrilled to welcome you to our organization's third monograph.

Unlike many organizations in which you may participate, the AMTE *Monograph Series*, *Connections Newsletter*, *Annual Conferences* and the sponsored section of *Contemporary Issues in Technology and Teacher Education* (CITE Journal) provide you with an opportunity to focus on your craft as a mathematics teacher educator — to examine and discuss current issues in mathematics teacher education and professional development and share related ideas and information.

The concept of a monograph series was first conceived at the AMTE Board meeting in Orlando, Florida on April 5, 2001 when Susan Beal, Tad Watanabe and Denisse Thompson asked the Board to consider developing a journal, sponsored by AMTE, that would discuss the practical issues of teacher education and applications of research. After a lively discussion, the Board suggested that the organization first consider generating a monograph series and gave Tad the charge of informally surveying the membership during the ATME reception.

A year later in Las Vegas, Nevada, he submitted a formal proposal to publish a monograph commemorating the 10th anniversary of the organization. Tad suggested that the proposed monograph should be "a collection of articles that address day-to-day practices of mathematics teacher educators." With the Board's vote, the monograph series was established.

From these initial steps to the present, the monograph series has been shaped and reshaped at each succeeding meeting of the Board. At the April 2003 meeting, the Board approved a proposal from East Carolina University (ECU), to publish the final product of its NSF supported MIDDLE MATH project to improve the undergraduate preparation of teachers of middle grades mathematics. The project facilitated the collaboration of mathematics and mathematics education faculty, providing a better sense of the changing content and pedagogical knowledge required to teach middle grades curricula reflective of the calls for reform, a greater knowledge of the experience of others who have set down this path, and an awareness of how the growing body of research on the teaching and learning of middle grades mathematics and undergraduate mathematics might impact a teacher preparation program. The publication and dissemination of this second monograph was supported through funding provided by the National

Science Foundation, the North Carolina Statewide Systemic Initiative and East Carolina University.

In 2004, the Board provided additional structure for the future of the monograph series. It was decided the odd numbered monographs would be general Conversations on Mathematics Teacher Education, in the spirit of the first monograph, and the even numbered monographs would each have a specific focus. The fourth monograph was selected to be a forum for mathematics teacher educators to discuss the ways in which they have used published cases to help preservice and/or practicing teachers develop their knowledge base for teaching (i.e., knowledge of content, pedagogy, and students as learners) and the capacity to reflect on and learn from teaching. The working title for the fourth monograph is *Cases in Mathematics Teacher Education: Tools for Developing Knowledge Needed for Teaching.* Susan Friel and Peg Smith were chosen as editors.

Perhaps the most important decision regarding the monograph series made at its 2004 Board Meeting in Philadelphia was the Board's recognition that a monograph series editor was needed to maintain quality and consistency throughout the series. The Board unanimously recruited and appointed Denisse R. Thompson to meet this challenge.

With the completion of the first monograph, the Board determined in 2005 that it was important to provide the monograph series to all current AMTE members. Rather than warehouse and sell copies, the series would be archived on the web in a "members-only" section. Members would be granted permission to download, copy and distribute monograph chapters for educational purposes.

In 2006 the Board voted to include the Monograph Series General Editor as a non-voting member at all Board meetings and on the Board's list serve. Under Denisse's guidance the Board established a timeline for monograph production and a structure for the selection of future editors and their editorial boards.

AMTE members are learning to anticipate a *Call for Manuscripts* each spring with a summer deadline. The call for manuscripts for the fifth monograph will be announced at the 2007 Annual Conference and will be posted on the web during the spring semester. The co-editors for this monograph are Fran Arbaugh and Mark Taylor.

The Board will be entertaining proposals for topics and recommendations for appropriate co-editors for the sixth monograph in time for appointments next spring.

Before closing, I would be remiss if I didn't acknowledge the efforts of the AMTE members who made this monograph possible. From across the country and representing all of the multifaceted dimensions of mathematics teacher education, twenty-two chapters were submitted. Space limitations forced the *Third AMTE Monograph* Editorial Board to select only nine of these for inclusion here. On behalf of the AMTE members, my thanks to the authors who cared enough to share their thoughts and experiences, to the hard working members of the editorial board who carefully reviewed, reflected and debated through the selection process, to the co-editors who worked with the authors to craft and shape the product you are holding, and finally to the Monograph Series General Editor, who assured that quality and integrity were maintained throughout the process.

Whether it's synchronously through the AMTE *Annual Conference*, or asynchronously through the *Connections Newsletter*, the sponsored section of the CITE Journal, or the AMTE *Monograph Series*, your AMTE membership provides *colleagues across the hall* to help you reflect on the issues that shape your professional life.

Sid Rachlin
AMTE President 2005–2007

Rider, R. L. and Lynch-Davis, K.
AMTE Monograph 3
The Work of Mathematics Teacher Educators
©2006, pp. 1-9

1

Continuing the Conversation on Mathematics Teacher Education

Robin L. Rider
East Carolina University

Kathleen Lynch-Davis
Appalachian State University

A major focus of teacher education is the development of preservice teachers. However, it should not be the only focus of those who work in teacher education. Educating inservice teachers is equally important and the conversation among those involved in mathematics teacher education needs to include discussion of this group as well. This conversation also highlights a need for professional development for teacher educators and research on the development of teacher educators. This paper discusses issues in educating all of these groups of individuals in an effort to continue the conversation among those involved in mathematics teacher education.

Professional development is an important aspect of the work in which mathematics educators engage. In this monograph, we broadly define professional development so that it captures the teaching and learning of preservice and inservice teachers and teacher educators. During the initial conception of this monograph, we thought it would provide an opportunity to renew the conversation about professional development between mathematics teacher educators that began in volume one of the AMTE monograph series (Watanabe & Thompson, 2004). After reviewing and selecting manuscripts, three subcategories of research emerged: teaching preservice teachers; teaching inservice teachers; and teaching teacher educators. These three areas, although having the common goal of creating good mathematics teachers, are significantly different to warrant individual treatment as separate research foci. This collection of articles is not meant to be exhaustive but a sharing of the work in which our colleagues are engaging.

Teaching Preservice Teachers

The four articles in this monograph that have implications for teaching preservice teachers address three concerns: increasing the mathematical content knowledge of preservice elementary teachers; gender equity; and capstone courses for preservice secondary teachers.

Lloyd (Chapter 2) and Flowers and Rubenstein (Chapter 3) outline ways to use mathematically rich problems from *Standards*-based K-12 curricula to enhance the content knowledge of preservice elementary teachers. The use of such problems helps preservice teachers examine curricula from the perspective of a student and also as a teacher and grounds preservice teachers in the mathematics they will teach while challenging their beliefs about traditional K-12 curriculum. Furthermore, the use of these curriculum materials has the potential to increase teachers' mathematical content knowledge consistent with the guidelines outlined in the *Principles and Standards for School Mathematics* (PSSM) (National Council of Teachers of Mathematics (NCTM), 2000) which states that "teachers must understand deeply the mathematics they are teaching." The Conference Board of the Mathematical Sciences (CBMS) (2000) has also recommended that mathematics content courses should be taught so they make connections with the school mathematics which teachers are expected to teach.

There are many inside and outside of education who subscribe to conventional wisdom that the majority of learning HOW to teach occurs when one actually starts teaching, and thus, teacher education programs are of little value. However, one overlooked fact is that all teachers have 16+ years of observations of teaching which shape their beliefs about teaching practices (Kennedy, 1999). These experiences impact preservice teachers' beliefs about mathematics and the way mathematics is taught effectively, perhaps explaining why the majority of teachers teach in the traditional manner in which they were taught. Currently, the usual course of study for educating preservice teachers is a series of college mathematics courses and then several mathematics education courses in which they learn how to apply that mathematics to what they are expected to teach in K-12 education. It is possible, in fact probable, that there is little or no connection between the mathematics courses or the mathematics education classes that typical preservice teachers take. Thus at the end of their college experience, mathematics may still appear to be a disconnected collection of topics which, in their view, has little or no connection to the school mathematics about which they are now expected to teach.

According to the Connections Standard in the PSSM, mathematics teachers should provide an instructional program which fosters understanding of the "interrelatedness of mathematical ideas" so that students learn the efficacy of mathematics as well as the mathematics itself. If teachers are taught in a disconnected fashion, it is unreasonable to assume that they will necessarily make those connections when they start teaching. Thus, the need for communication between mathematicians and mathematics educators is critical in the education of preservice teachers to foster ideas such as using rich content problems and non-traditional teaching practices.

Many preservice teachers operate under the belief that the way they learned mathematics is an effective way to teach mathematics to their students, that is, it worked for them so it should work for everyone. This belief needs to be challenged in preservice teacher education courses. In both K-12 and college classes, the majority of teachers were taught in traditional mathematics courses, with few connections between different mathematics strands or to other subject areas. There are also few teachers who were taught with reform teaching methodologies or from *Standards*-based practices. Teacher educators and policy makers cannot expect preservice teachers to implement *Standards*-based curricula when they have never been exposed to them or to use pedagogical strategies that would enhance mathematical understanding for all students when they have limited experience with successful mathematics teaching practices. Hence, it is critical for mathematics teacher educators to converse about strategies for teaching preservice teachers pedagogy which fosters mathematical understanding for all students.

Pedagogical content knowledge (PCK) for mathematics education consists of knowing what mathematics to teach, how to teach it, and why it should be taught. The CBMS recommends a capstone course to enhance preservice teachers' PCK. In the early years of teaching, new teachers rarely consider what mathematics is important to teach, but let it be determined for them by outside sources, such as a prescribed textbook, pacing guide, or state mandated course of study. Education programs need to help preservice teachers consider what mathematics is important to teach, why it is important, and how it connects to mandated curricula.

Instituting a capstone course in teacher preparation programs is one strategy to provide preservice teachers with an opportunity to examine connections between mathematical ideas in the K-12 curriculum that they will teach. Developing a capstone course is not an easy task however. Decisions, such as what content to include in such a course, how to structure the course, and how the course will enhance PCK and preservice teachers' views of mathematics as a connected whole, are critical discussion topics among those involved in mathematics teacher education. Teacher educators who have developed or are revising such a course are an invaluable asset in leading or facilitating such discussion. Loe and Rezac (Chapter 4) provide a glimpse of a capstone course for secondary teachers that hopes to bridge this divide.

Mathematics education professionals also must attend to preservice teachers' ability to teach the mathematics that they know equitably to all students. Even with previous strides to resolve issues of racial and gender equity in K-12 education, those issues still exist, as evidenced by the achievement gap on mandated tests by underserved populations. Thus, the conversation about racial and gender equity, as addressed by Breyfogle and Kress (Chapter 5), needs to be revisited by mathematics teacher educators.

There are many issues which shape preservice teacher education beyond what is possible to present in any one volume. The intent here is to reenergize the conversation among those involved in the education of future K-12 mathematics teachers. This sharing of ideas and best practices can enhance the mathematical and pedagogical education of mathematics teachers and ultimately the education of K-12 mathematics students.

Teaching Inservice Teachers

The three articles in the monograph that address teaching inservice teachers center around creating communities that provide high quality professional development for both novice and experienced mathematics teachers. Creating such communities is often a difficult process given the increasing demands on K-12 teachers.

The transition from preservice teacher to inservice teacher is often a challenge, even for well-prepared novice teachers. This is evidenced by attrition rates, especially in the areas of mathematics and science, which are highest for teachers leaving the profession in the first five years of teaching (Ingersoll, 2001). When preservice teachers enter their first classrooms, they are frequently overwhelmed by a myriad of daily details dealing with classroom organization and management and administrative responsibilities. As these fledgling teachers struggle for survival in the classroom, the textbook becomes their lifeline and pedagogy is reduced to lecture, drill and practice. To expect inexperienced teachers to break from a traditional emphasis and to institute reform methods is an idealistic notion when viewed from the perspective of the day-to-day realities of their classrooms. Once these teachers have gained experience, they are entrenched in this traditional teaching mode and rely heavily on the adopted text as their source of determining the mathematics they will teach. Mathematics educators need to continue the conversation regarding the implementation of *Standards*-based curricula and reform methodologies in both new and experienced teachers' classrooms.

One possible way to smooth the transition to teaching, retain quality teachers in the classroom, and encourage new teachers to institute reform methods is by providing mentoring for new teachers. Although mentoring might be easy in large schools with an abundance of qualified mathematics teachers, it is more challenging in rural areas where a new teacher might be the only mathematics teacher for a grade band, as evidenced by the research presented by Luebeck (Chapter 6). Finding an available qualified mentor who can answer a new teacher's questions regarding content and pedagogical aspects of their specific teaching assignment can be a problem in rural areas. Although a mentor may give novice teachers classroom management advice, the mentor may not be able to offer guidance about mathematics because the mentor teaches a different subject area. Often in rural settings, teachers feel isolated and disconnected from their subject area peers (Storer & Crosswait, 1995). Technology, as described by Luebeck (Chapter 6), may provide a solution to this mentoring problem through the use of internet media which provides two-way communication between rural teachers and mentors in other schools, districts, or even universities.

Given the availability of web based media, such as email, instant messaging, weblogs, and high-end video conferencing technologies, teachers and teacher educators have more ways to stay connected than ever before. The question remains of how to utilize these technologies to connect practicing teachers and teacher educators in appropriate ways to enhance professional development of all involved. Although there has been some research on best practices in distance education and distance professional development (Rider & Manning, 2005), there is a lack of understanding of how these media could assist the mentoring process of new teachers.

Establishing communities of practice (Wenger, 1998) among teachers, teacher educators, and other education professionals to look at standards, curriculum, classroom practice, student understanding, and rich problems in mathematics has been shown to be an effective professional development activity for both new and experienced teachers. Crespo and Featherstone (Chapter 7) address ways that they have created teacher groups and used rich mathematical problems to provide professional development, thus developing a community of practice. These communities of practice give teachers support from their peers as they grapple with issues of mathematical content, strategies that promote conceptual understanding for their students, diagnosing problems in students' work, connecting research to practice, and pedagogical aspects of teaching. All are issues when implementing reform curricula. Silver, Mills, Castro, and Ghousseini (Chapter 8) address similar issues using a modified Japanese lesson study model with case analysis and discussion.

Teachers in small schools and rural communities do not have as many opportunities for professional development as do teachers in larger areas. Although it is possible to form communities of educators within a single school (Crespo & Featherstone, 2003, 2002, 2001), distance learning technology has been used to maintain communities of practice that are formed during teacher professional development activities with teachers from different schools and even different areas of the country (Rider & Manning, 2005). For many teachers, especially those in rural areas, technology becomes a key factor in making it possible for them to engage in these professional development activities, allowing them to explore problems in mathematics, observe and share classroom experiences, and examine student understanding (Rider & Hunting, 2006). The use of technology opens up a wealth of opportunities for teachers and a range of instructional design issues for teacher educators. Teacher educators need to consider and discuss how to utilize technological tools in ways that foster the same learning as face-to-face meetings.

The three papers by Luebeck, Crespo and Featherstone, and Silver et al. begin to touch on issues of creating community in professional development. Hence, they are first steps in meeting a critical need for conversation among teacher educators on how to create these communities and foster growth for participating teachers as they implement *Standards*-based curricula.

Teaching Teacher Educators

The final two papers in the monograph represent a growing concern and research area often overlooked in teacher education, namely the education and professional development of the teacher educator. Teacher educators come from diverse educational backgrounds and they may or may not have had any instruction in preservice or inservice teacher education. As evidenced by Van Zoest, Moore, and Stockero (Chapter 9) and Sztajn, Ball, and McMahon (Chapter 10), even experienced teachers may have difficulty transitioning from being a mathematics teacher to a mathematics teacher educator. It is imperative that the profession start and continue a conversation on how to educate this group of individuals.

Preparing teacher educators to provide professional development and meaningful learning opportunities to both preservice and inservice mathematics teachers is vital to the creation of better mathematics teachers. The recognition that there is a specialized body of knowledge and experiences that help the mathematics educator prepare preservice and inservice teachers is of great importance. Often a mathematics education Ph.D. or mathematics Ph.D. has experiences and coursework on preservice teacher education throughout the program but there has never been explicit attention on how to transition from a Ph.D. student into a teacher educator. To some, this might seem to be an obvious transition; similar to preservice teachers transitioning into a classroom teaching experience, the transition can be challenging for new teacher educators. The two articles featured in this monograph may help the mathematics education community think about experiences which may be beneficial in the development of teacher educators.

Conclusion

In order to strengthen the field of mathematics education, the conversation among mathematicians, mathematics teacher educators, and other professionals involved in the education of mathematics teachers needs to continue and increase. In too many cases, there is little connection between departments of mathematics, mathematics education, and outside professional developers. Collectively, these individuals need to work together to produce not only more mathematics teachers but better prepared and trained teachers. The critical shortage of qualified mathematics teachers in K-12 (Sterling, 2004) and in tertiary (Reys, 2006) mathematics education positions makes the need for this conversation and collaboration more important than ever.

References

Conference Board of the Mathematical Sciences. (2000). *Mathematical Education of Teachers*. Washington, DC: Author.

Crespo, S., & Featherstone, H. (2003; 2002; 2001). *Communities of practice to improve mathematics and mathematics teaching: Years 1-3.* Technical Reports to the Lucent Technologies Foundation. East Lansing, MI: Michigan State University.

Ingersoll, R. (2001). Teacher turnover and teacher shortages: An organizational analysis. *American Educational Research Journal, 38*(3), 499-534.

Kennedy, M. (1999). The role of preservice teacher education. In L. Darling-Hammond & G. Sykes (Eds.), *Teaching as the learning profession: A handbook of policy and practice* (pp. 54-85). San Francisco: Jossey-Bass.

National Council of Teachers of Mathematics. (NCTM) (2000). *Principles and standards for school mathematics.* Reston, VA: Author.

Reys, R. E. (2006). A report of jobs in mathematics education in institutions of higher education. *Journal for Research in Mathematics Education, 36*(4), 262-269.

Rider, R. L., & Hunting, R. (2006). The VideoPaper: Issues in implementation of a multimedia tool for professional self-dialogue and communication in mathematics education. In P. Grootenboer, R. Zevenbergen & M. Chinnappan (Eds.), *Identities Cultures and Learning Spaces, Vol 2* (pp. 440-446). Canberra: MERGA Inc.

Rider, R. L., & Manning, D. D. (2005). Defining best practices in distance professional development: Lessons learned from a national videoconferencing project. In V. Uskov (Ed.), *International Conference on Computers and Advanced Technology in Education* (pp. 498-502). Oranjestad, Aruba: ACTA Press.

Sterling, D. R. (2004). The teacher shortage: National trends for science and mathematics teachers. *The Journal of Mathematics and Science: Collaborative Explorations, 7,* 85-96. Available online <http://www.math.vcu.edu/g1/journal/Journal7/Part%20I/Sterling.html> Accessed July 1, 2006.

Storer, J. H., & Crosswait, D. J. (1995). Delivering staff development to the small rural school. *Rural Special Education Quarterly, 14*(3), 23-30.

Watanabe, T., & Thompson, D. R. (Eds). *The work of mathematics teacher educators: Exchanging ideas for effective practice.* AMTE Monograph Series, Volume 1. San Diego, CA: Association of Mathematics Teacher Educators.

Wenger, E. (1998). *Communities of practice. Learning, meaning, and identity*. Cambridge, MA: Cambridge University Press.

Robin L. Rider, a Nationally Board Certified AYA Mathematics teacher, is currently an Assistant Professor in the Department of Mathematics and Science Education at East Carolina University. She received her Ph.D. from North Carolina State University and teaches undergraduate and graduate courses in mathematics education. Her research interests include using technology in K-12 mathematics classrooms to enhance student understanding, using multiple representations when teaching algebra to diverse learners, and statistics education.

Kathleen Lynch-Davis, a former middle school mathematics teacher, is currently an Assistant Professor in the Department of Curriculum and Instruction at Appalachian State University. Dr. Lynch-Davis received her B.S. and M.A. Ed. at the University of North Carolina at Wilmington and her Ph.D. at Indiana University. She currently teaches mathematics education and curriculum courses and also works with K-8 mathematics teachers in the Boone, NC area. Her research interests include communication in mathematics, particularly the role of writing in mathematics, teaching and learning geometry, and meeting the needs of diverse learners in elementary and middle grades mathematics classrooms.

Lloyd, G. M.
AMTE Monograph 3
The Work of Mathematics Teacher Educators
©2006, pp. 11-27

2

Using K-12 Mathematics Curriculum Materials in Teacher Education: Rationale, Strategies, and Preservice Teachers' Experiences[1]

Gwendolyn M. Lloyd
Virginia Polytechnic Institute and State University

Engagement with K-12 curriculum materials has long been utilized as a successful professional development strategy for inservice teachers, but only recently has it emerged as a valuable preservice teacher education activity. With the availability of Standards-based mathematics curriculum materials, teacher educators have unique opportunities to challenge preservice teachers' conceptions of mathematics teaching and learning through the use and analysis of curriculum materials. This paper provides a rationale and proposes strategies for adapting K-12 curriculum materials for preservice mathematics teacher education, and describes preservice elementary teachers' experiences learning with middle-school curriculum materials in undergraduate mathematics courses.

Teachers engaged in current mathematics education reforms attempt to establish classrooms in which students engage actively and cooperatively in exploration and discussion to solve rich problems and reason mathematically (National Council of Teachers of Mathematics [NCTM], 1989, 2000). To support the vision of mathematics instruction presented in the 1989 *Standards* document, a dozen or more federally-funded projects developed novel curriculum materials during the 1990s. Students' problem-solving abilities and conceptual understandings of mathematics appear to be impacted positively by the use of these materials (Senk & Thompson, 2003). Because they are a

[1] The work reported in this paper was funded in part by the National Science Foundation under Award Number 9983393 (*Building a Theory of Teacher Learning With and About Mathematics Curriculum: The Role of Innovative K-12 Materials in Elementary Teacher Education*). Any opinions, findings, conclusions or recommendations expressed in this publication are those of the author and do not necessarily reflect the views of the National Science Foundation.

source of mathematically rich problems and instructional activities, these materials may also give rise to potent opportunities for *teachers* to learn (Ball & Cohen, 1996). Although many reports have illustrated the challenges faced by teachers when implementing *Standards*-based materials (Frykholm, 2004; Lambdin & Preston, 1995; Lloyd, 1999; Manoucheri & Goodman, 1998; Wilson & Lloyd, 2000), research also has offered images of teachers learning about mathematics and pedagogy while using these materials (Lloyd, 2002a; Remillard, 2000; Van Zoest & Bohl, 2002).

The challenges of curriculum implementation, together with the potential of curriculum materials to promote and support teacher learning, suggest that preservice teachers may benefit from early experiences with *Standards*-based K-12 curriculum materials. Engaging teachers with textbooks and curriculum materials is a commonly used and effective professional development strategy for inservice teachers (Loucks-Horsley, Love, Stiles, Mundry, & Hewson, 2003). Increasingly, teacher educators are using K-12 curriculum materials, particularly *Standards*-based materials, in the preparation of preservice teachers (Frykholm, 2005; Lloyd, 2002b; Lloyd & Behm, 2005; Tarr & Papick, 2004). This paper provides a rationale for using such materials in preservice mathematics teacher education, proposes strategies for using mathematics curriculum materials in courses for preservice teachers, and describes preservice elementary teachers' experiences using middle-school curriculum materials in undergraduate mathematics courses.

Rationale for the Use of Curriculum Materials in Teacher Education

Many preservice teachers possess weak knowledge and narrow views of mathematics and mathematics pedagogy. Because such conceptions deeply affect the learning-to-teach process, teacher educators are faced with the task of creating opportunities for preservice teachers to develop useful, dynamic conceptions of mathematics and pedagogy. As described in the following sections, experiences with curriculum materials have potential to challenge preservice teachers' conceptions of mathematics, teaching and learning, and curriculum so that learning (and re-learning) in these areas may occur.

Learning Mathematics

A primary aim of engaging preservice teachers with K-12 curriculum materials is to facilitate the learning or re-learning of the school mathematics curriculum. For preservice teachers, this learning involves revisiting mathematical ideas to include conceptual or relational understandings (Skemp, 1987), as well as exploring unfamiliar curricular areas, such as probability, statistics, and discrete mathematics, that are integral to the *Standards'* vision (NCTM, 1989, 2000). Preservice teachers' views of mathematical activity also need to be developed so that these views include *mathematics* as a dynamic, evolving body of knowledge (Ernest, 1991) and *understanding* as the capacity to use mathematics to reason, to communicate, and to pose and solve meaningful problems (Hiebert et al., 1996).

Textbooks that have been written for mathematics courses for preservice teachers tend to be traditional in design, essentially "telling" preservice teachers what they need to know. In contrast, reform-oriented curriculum materials contain diverse problems and activities that challenge preservice teachers and require them to develop and test their own mathematical ideas in the social setting of the classroom. Use of curriculum materials has potential to offer preservice teachers experiences "in doing mathematics—in exploring, guessing, testing, estimating, arguing, and proving . . . They should learn mathematics in a manner that encourages active engagement with mathematical ideas" (Mathematical Sciences Education Board & National Research Council, 1989, p. 65). As preservice teachers revisit mathematical content from new perspectives, they begin to translate the knowledge developed as students of mathematics into pedagogical content knowledge - knowledge of mathematics *for teaching* (Shulman, 1987).

Learning about Mathematics Teaching and Learning

Another purpose of using curriculum materials in teacher education is to promote the view that mathematics is learned through an active, social process of construction rather than through transmission (Cobb, 1995; Davis & Maher, 1990; von Glasersfeld, 1984). As preservice teachers engage in mathematical activities presented in curriculum materials, examination of their own learning processes may help them recognize the significance of learning that can occur during inquiry-based, student-centered activities. Research has indicated that a teacher's views about how students engage in mathematical activity and learn may greatly inhibit (or support) inquiry-based instruction (Fennema et al., 1996). In many cases, teachers' self-efficacy is rooted in models of learning that are not consistent with current reforms – teachers have traditionally felt most self-efficacious when they tell students what they need to know (Smith, 1996). Preservice teachers' experiences with curriculum materials encourage them to explore, conjecture, and reason about mathematical problems and situations – experiences that have potential to empower them as learners and teachers of mathematics.

Learning about Curriculum

Preservice teachers can also undertake critical analyses of the mathematical content and instructional designs of curriculum materials. Through analysis of mathematics curriculum materials, preservice teachers have opportunities to develop sophisticated views of curriculum (Ball & Cohen, 1996; Ben-Peretz, 1990; Lloyd & Behm, 2005). Ben-Peretz (1990) contends that curriculum "elaboration and change may be considered a cardinal component of [teachers'] professional activities" (p. 110) and suggests that particular curricular capabilities and understandings are needed. Analysis of issues that underlie curriculum development, termed "choice points" by Connelly (1972), can make mathematical and instructional features of curriculum materials apparent to preservice teachers. For example, it is important for preservice teachers to recognize that there are many ways that curriculum developers might approach instruction about a particular mathematical topic. Embedded in such approaches are different philosophies of teaching and learning that fulfill a variety of cultural and social purposes. Experiences with curriculum analysis can prepare preservice teachers to make reasoned decisions about the selection and adaptation of instructional materials in their future mathematics classrooms.

Strategies for Using Curriculum Materials in Teacher Education

Reform-oriented curriculum materials can be used in a variety of ways and for a variety of purposes in mathematics and methods courses for teachers.[i] Appendix A presents a selection of strategies for using curriculum materials for different purposes in preservice teacher education. Perhaps the most natural way to use curriculum materials is to engage preservice teachers in exploring mathematical problems and activities with the goal of learning or re-learning subject matter. Although curriculum designers wrote these materials for K-12 students, the mathematics in the materials can be challenging, even for preservice teachers. Many aspects of the content are unfamiliar to preservice teachers, both because conceptual aspects of mathematics are emphasized and because some of the content is relatively new to the school curriculum (e.g., probability, statistics, and discrete mathematics).

Because mathematical ideas are explored in reform-oriented ways in these curriculum materials (for instance, in real-world contexts, through small group discussions, with graphing calculators), many teaching and learning issues can be explored. For instance, work with teachers' editions (which include descriptions of possible student responses or work) to plan instructional activities can help preservice teachers think about the processes through which students develop understandings during particular classroom activities. Due to the rich mathematical content and innovative instructional design of the materials, opportunities for the integration of content and pedagogy abound (Ball, 2000). In fact, many of the strategies in Appendix A can be used to accomplish multiple purposes related to mathematics, teaching and learning, and curriculum.

Adapting K-12 Materials for Use by Preservice Teachers

As teacher educators think about using curriculum materials with preservice teachers, it is important to attend carefully to the selection and adaptation of materials. Tasks and activities originally written for K-12 students must be selected and adapted in ways that are relevant and useful to preservice teachers. Mathematics problems that provide learning opportunities for children are not necessarily productive or educative problems for consideration by teachers. The greatest potential for teachers' learning about mathematics is through engagement with tasks that are mathematically challenging, emphasize conceptual understandings, address common misconceptions, and have potential to illustrate connections among concepts, representations, and real-world contexts. Tasks with greatest potential to illustrate or question important teaching and learning issues are those with multiple solution strategies, technology use, meaningful real-world contexts, problem-based learning, experimentation, and investigation.

Once tasks or activities have been selected, adaptations must be made to suit these materials to *teacher* learning. Adaptations must go beyond preservice teachers' use of the materials *as students* but also must include preservice teachers' engagement with the materials *from a teaching perspective.* For example, a high school student would not typically study the same mathematical concept using two different curriculum units, but preservice teachers might (e.g., for comparison purposes). Uses of curriculum materials in preservice teacher education, such as those listed in Appendix A, are likely to be most productive if instruction begins with a focus on the mathematics of the curriculum materials (Lloyd & Behm, 2005; see also Hill & Ball, 2004) so that preservice teachers' pedagogical views and decisions can be firmly based in knowledge and reasoning about the mathematics they will be teaching.

An Example of a Curriculum-Based Strategy

A detailed illustration from my own use of K-12 curriculum materials with preservice teachers in mathematics courses over the past eight years may be helpful to other teacher educators. Selected units from several middle-school curriculum projects comprise the main texts for these courses.[ii] Preservice teachers spend extended periods of time in class using the curriculum units to learn or re-learn mathematical subject matter (number concepts, probability and statistics, discrete mathematics, and geometry) and consider related pedagogical issues.

The example shared in this section involves two middle school units, *Prime Time* (Lappan, Fey, Fitzgerald, Friel, & Phillips, 1996) and *Reflections on Number* (Mathematics in Context [MiC], 1998), which both present innovative activities related to number-theoretic ideas such as factors, multiples, divisors, and primes. Although the books contain distinctly different activities, they have one major mathematics problem in common. The Locker Problem from *Prime Time* and the Changing Positions Problem from *Reflections on Number* involve the idea that perfect squares are the only numbers with an odd number of factors. The Locker Problem reads,

> There are 1000 lockers in the long hall of Westfalls High. ... The lockers are numbered from 1 to 1000. . . . Student 1 runs down the row of lockers and opens every door. Student 2 closes the doors of lockers 2, 4, 6, 8, and so on to the end of the line. Student 3 changes the state of the doors of lockers 3, 6, 9, 12, and so on to the end of the line. (The student opens the door if it is closed and closes the door if it is open.) Student 4 changes the state of the doors of lockers 4, 8, 12, 16, and so on . . . until all 1000 students have had a turn. When all the students are finished, which lockers are open? (pp. 58-60)

In the Changing Positions Problem in *Reflections on Number*, numbered students stand up and sit down in a way similar to lockers being opened and closed in the Locker Problem. Although these problems are similar and are presented as key problems in the two curriculum units, the problems play different roles in the development of mathematical ideas in the two units. In *Prime Time*, the Locker Problem appears at the end of the unit after students have developed the mathematical ideas needed to solve it. In contrast, the Changing Positions Problem appears early in *Reflections on Number* as an initial experience with the main mathematical ideas that are explored in the unit.

These problems (and the curriculum units containing them) provide excellent contexts for preservice teachers to develop new mathematical and pedagogical understandings. The problems are challenging, demand that learners develop and apply conceptual understandings, have interesting real-world contexts, and involve important concepts and representations of those concepts. After the preservice teachers in my mathematics course work on most of the problems and activities in *Prime Time* and *Reflections on Number*, they reflect on their learning and analyze the curriculum materials. Appendix B offers some of the questions about the Locker Problem and the Changing Positions Problem to which the preservice teachers respond, as well as examples of two preservice teachers' responses. As Annika and Joy's[iii] responses suggest, curriculum analysis can follow naturally from preservice teachers' engagement with the mathematics problems in the curriculum materials. Distinctions between different reform-oriented curricula (and similarly, between reform and traditional curricula) provide rich opportunities for preservice teachers to explore, and possibly experience, multiple approaches to mathematical subject matter and mathematics pedagogy.

Preservice Elementary Teachers' Experiences

This section supplements the strategies presented in the previous section by discussing preservice elementary teachers' experiences using *Standards*-based middle school curriculum materials. Most preservice elementary teachers are surprised to find that working with the middle-school curriculum materials as learners of mathematics can be challenging. Preservice teachers quickly become aware that the curriculum materials are different from the mathematics materials and textbooks they have used in the past. They are impressed with the curriculum materials' focus on multiple solutions to problems and emphasis on why particular methods and procedures work and make sense. For example, as preservice teachers attempt to determine how lattice multiplication works, they also think about the role of partial products in conventional multiplication algorithms and why those multiplication algorithms make sense.

One preservice teacher (Meg) expressed differences that she observed between the middle-school curriculum materials and more traditional mathematics textbooks:

> A typical textbook has a lesson and a group of questions. What we have is similar but it doesn't teach what we're learning. . . It's more questions and a lot of explanations. They'll ask questions and then "Why? Why this?" or "What could be different?"

The unfamiliar format of the curriculum materials can initially present frustration for some preservice teachers who desire more guidance from a textbook. As Jessica expressed,

> I like to have an explanation and then maybe an example or two and then some problems. . . . Some of the books just say "do this problem" and you don't really have any basis for what you're doing, or where to start.

Over time, most preservice teachers come to appreciate the place of investigation and discussion in the development of mathematical understanding. To varying degrees, preservice teachers' levels of comfort appear to increase as they accumulate experiences solving challenging mathematics problems from the materials, particularly with the support and collaboration of classmates.

Preservice teachers also seem to benefit from course activities that draw their attention to pedagogical issues through reflection on and analysis of the mathematical work they are doing. However, opportunities for learning about pedagogy and curriculum do not occur only during course assignments that explicitly focus on those areas. During class discussions about mathematical questions and ideas, preservice teachers themselves frequently raise issues related to students' learning through the activities presented in the curriculum materials, the design or philosophy of the materials, and ways that teachers would use the materials (or similar materials) with children.

Consider a class discussion that followed preservice teachers' work on an investigation in *Reflections on Number* (MiC, 1998) in which the teachers had to assume the role of a tutor who helps a child, Harvey, with a mathematics problem related to division with zero. The investigation begins,

> During one tutoring session, you say to Harvey, "For every multiplication problem, there are related division problems. For example, you can take $3 \times 7 = 21$ and write $21 \div 3 = 7$." . . . Next, you ask Harvey to write a division statement for the following situation: "You have six stickers, and you share them with no one." Harvey smiles, "I know that! I have six stickers and I share with nobody. The division statement is $6 \div 0 = 6$. I am sharing with nobody, so I have all six stickers." (p. 28)

During our class discussion about division involving zero, the preservice teachers commented extensively about the role of the tutor (and the text, to a lesser extent) in teaching about this idea:

> *Beth*: I just want to say that I was personally offended that I was being a tutor but they didn't explain the problem to the kid right. Wasn't it something about sharing something with only himself? And he says if you divided it with no one? If you didn't share it with anyone, then I, the tutor, would have been like, "Well, don't forget yourself – you're *one*." And then there would have been none of this confusion. You're not dividing by zero because you're *one*, so it's six divided by one, which is six! Why didn't he [the tutor] just tell him [the student]? (Sighs, class laughs)
>
> *Instructor*: Well, what do you think of the tutor in the problem? (to the rest of the class)
>
> *Several preservice teachers*: Bad. Weak. (Some teachers shrug their shoulders.)
>
> *Nicole*: The tutor didn't explain. . . . There was no explanation at all that I could find in that (pointing at the *Reflections on Number* book on her table).

After the class had shifted attention away from the tutor and back to the idea of "6 divided by 0," Jessica introduced some additional comments that turned the discussion again to Beth's idea that the tutor should have told Harvey that he had forgotten to include himself.

> *Jessica*: To put myself in a kid's point of view, I would think that six divided by zero is six. You have six, you divide it by nothing, you have six. I think Harvey's explanation for his age made perfect sense and I was like, "Well how do you say it's not?" And the tutor used a calculator. That's not a great learning tool. It still didn't tell him *why*. He [the tutor] is just like, "Oh, it doesn't work. The calculator says it doesn't work so it doesn't work."
>
> *Cathy*: Six times zero would be zero and not six. I mean, it only checks one way, which may confuse them, like she [Jessica] said. Explaining would help!
>
> *Beth*: She [the tutor] should have said, "Well, you're splitting it by no one but you have it yourself, so it's really six divided by one."

In this short excerpt, preservice teachers made comments about a wide range of mathematical and pedagogical issues, including the lack of direct explanation by the tutor in the investigation, appropriate use of calculators in teaching and learning, mathematical considerations related to division and its inverse, and ways for teachers to approach children's learning. Conversations like this one offer teacher educators invaluable opportunities to elicit and discuss teachers' struggles and questions about teaching and learning that relate to mathematical work with curriculum materials.

Conclusions

Given the demands of the current reform movement in mathematics education, teacher educators should consider seriously the powerful role that *Standards*-based curriculum materials can play in the preparation of future teachers. Experiences with innovative curriculum materials may compel preservice teachers to recognize the significance of the learning that can occur during inquiry and student-centered activities. Moreover, explicit emphasis in teacher education may enable beginning teachers to use textbooks, curriculum materials, and other resource materials to teach themselves and their students. If teachers can learn to use their textbooks for their own personal development, then they may be prepared to learn from and deal productively with the types of materials that will continue to emerge in school settings in the future.

References

Ball, D. L. (2000). Bridging practices: Intertwining content and pedagogy in teaching and learning to teach. *Journal of Teacher Education, 51*, 241-247.

Ball, D. L., & Cohen, D. K. (1996). Reform by the book: What is - or might be - the role of curriculum materials in teacher learning and instructional reform? *Educational Researcher*, *25*(9), 6-8, 14.

Ben-Peretz, M. (1990). *The teacher-curriculum encounter: Freeing teachers from the tyranny of texts*. Albany: State University of New York Press.

Cobb, P. (1995). Where is the mind? Constructivist and sociocultural perspectives on mathematical development. *Educational Researcher, 23*(7), 13-20.

Connelly, F. M. (1972). The functions of curriculum development. *Interchange, 2*(3), 161-177.

Davis, R. B., & Maher, C. A. (1990). The nature of mathematics: What do we do when we do mathematics? In R. B. Davis, C. A. Maher, & N. Noddings (Eds.), *Constructivist views on the teaching and learning of mathematics* (pp. 65-78). Reston, VA: National Council of Teachers of Mathematics.

Ernest, P. (1991). *The philosophy of mathematics education.* Hampshire, UK: Falmer.

Fennema, E., Carpenter, T. P., Franke, M. L., Levi, L., Jacobs, V. R., & Empson, S. B. (1996). A longitudinal study of learning to use children's thinking in mathematics instruction. *Journal for Research in Mathematics Education, 27*(4), 403-434.

Frykholm, J. A. (2004). Teachers' tolerance for discomfort: Implications for curricular reform in mathematics. *Journal of Curriculum and Supervision, 19*(2), 125-149.

Frykholm, J. A. (2005). Innovative curricula: Catalysts for reform in mathematics teacher education. *Action in Teacher Education, 26*(4), 20-36.

Hiebert, J., Carpenter, T. P., Fennema, E., Fuson, K., Human, P., Murray, H., et al. (1996). Problem solving as a basis for reform in curriculum and instruction: The case of mathematics. *Educational Researcher, 25*(4), 12-21.

Hill, H. C., & Ball, D. L. (2004). Learning mathematics for teaching: Results from California's mathematics professional development institutes. *Journal for Research in Mathematics Education, 35*, 330-351.

Lambdin, D. V., & Preston, R. V. (1995). Caricatures in innovation: Teacher adaptation to an investigation-oriented middle school mathematics curriculum. *Journal of Teacher Education, 46*, 130-140.

Lappan, G., Fey, J. T., Fitzgerald, W. M., Friel, S. N., & Phillips, E. D. (1996). *Prime Time.* Palo Alto, CA: Dale Seymour.

Lloyd, G. M. (1999). Two teachers' conceptions of a reform curriculum: Implications for mathematics teacher development. *Journal of Mathematics Teacher Education, 2*, 227-252.

Lloyd, G. M. (2002a). Reform-oriented curriculum implementation as a context for teacher development: An illustration from one mathematics teacher's experience. *The Professional Educator, 24*(2), 51-61.

Lloyd, G. M. (2002b). Mathematics teachers' beliefs and experiences with innovative curriculum materials: The role of curriculum in teacher development. In G. Leder, E. Pehkonen, & G. Törner (Eds.), *Beliefs: A hidden variable in mathematics education?* (pp. 149-159). Utrecht: Kluwer.

Lloyd, G. M., & Behm, S. L. (2005). Preservice elementary teachers' analysis of mathematics instructional materials. *Action in Teacher Education, 26*(4), 48-62.

Loucks-Horsley, S., Love, N., Stiles, K. E., Mundry, S., & Hewson, P. W. (2003). *Designing professional development for teachers of science and mathematics* (2nd ed.). Thousand Oaks, CA: Corwin.

Manouchehri, A., & Goodman, T. (1998). Mathematics curriculum reform and teachers: Understanding the connections. *Journal of Educational Research, 92*(1), 27-41.

Mathematical Sciences Education Board & National Research Council. (1989). *Everybody counts: A report to the nation on the future of mathematics education.* Washington DC: National Academy Press.

Mathematics in Context. (1998). *Reflections on number.* Chicago: Encyclopaedia Britannica.

National Council of Teachers of Mathematics. (1989). *Curriculum and evaluation standards for school mathematics.* Reston, VA: Author.

National Council of Teachers of Mathematics. (2000). *Principles and standards for school mathematics*. Reston, VA: Author.

Remillard, J. T. (2000). Can curriculum materials support teachers' learning? Two fourth-grade teachers' use of a new mathematics text. *Elementary School Journal, 100*(4), 331-350.

Senk, S. L., & Thompson, D. R. (Eds.). (2003). *Standards-based school mathematics curricula: What are they? What do students learn?* Mahwah, NJ: Lawrence Erlbaum.

Shulman, L. S. (1987). Knowledge and teaching: Foundations of the new reform. *Harvard Educational Review, 57*(1), 1-22.

Skemp, R. R. (1987). *The psychology of learning mathematics.* Hillsdale, NJ: Lawrence Erlbaum.

Smith, J. P. (1996). Efficacy and teaching mathematics by telling: A challenge for reform. *Journal for Research in Mathematics Education, 27*(4), 387-402.

Tarr, J. E., & Papick, I. J. (2004). Collaborative efforts to improve the mathematical preparation of middle grades mathematics teachers. In T. Watanabe & D. R. Thompson (Eds.), *The work of mathematics teacher educators: Exchanging ideas for effective practice (AMTE Monograph 1)* (pp. 19-34). San Diego: Association of Mathematics Teacher Educators.

Van Zoest, L. R., & Bohl, J. V. (2002). The role of reform curricular materials in an internship: The case of Alice and Gregory. *Journal of Mathematics Teacher Education, 5*, 265-288.

von Glasersfeld, E. (1984). An introduction to radical constructivism. In P. Watzlawick (Ed.), *The invented reality* (pp. 17-40). New York: Norton.

Wilson, M. R., & Lloyd, G. M. (2000). The challenge to share mathematical authority with students: High school teachers reforming classroom roles. *Journal of Curriculum and Supervision, 15*, 146-169.

---

[i] This manuscript focuses on the incorporation of curriculum materials into teacher education coursework. Another promising context for preservice teachers' learning with curriculum materials, beyond the scope of this chapter, is student-teaching placements in the classrooms of teachers implementing innovative materials (see Van Zoest & Bohl, 2002).

[ii] Selected curriculum units correspond to the mathematical emphases typical to college mathematics textbooks for preservice elementary teachers.

[iii] All names of preservice teachers are pseudonyms. Unless otherwise noted, quotes from preservice teachers are taken from transcripts of interviews conducted outside of class.

Gwen Lloyd, Associate Professor in the Department of Mathematics at Virginia Tech, completed her doctorate in mathematics education in 1996 at the University of Michigan. She teaches mathematics courses for preservice elementary and secondary teachers and she received Virginia Tech's Alumni Award for Teaching Excellence in 2003. Gwen is currently the Principal Investigator of an NSF-funded study of preservice secondary teachers' interactions with curriculum materials in undergraduate mathematics courses. She is also a member of the Editorial Panel of *Journal for Research in Mathematics Education* and a Research Associate of the NSF-funded *Center for the Study of Mathematics Curriculum.*

Appendix A
Sample Strategies for Using Curriculum Materials [CMs]

For Learning About …	**Sample Strategies**
Mathematics	Solve mathematics problems and conduct particular experiments and investigations from selected CMs. Compare and contrast different methods for solving similar problems and analyze why different methods make sense. Create concept maps to identify connections among the different mathematical ideas explored in a particular unit. Compare representations and definitions of particular mathematical ideas (e.g., fractions, functions) in reform-oriented CMs and more traditional textbooks, with analysis of the mathematical and pedagogical implications of the differences.
Teaching and Learning Mathematics	Trace the development of one mathematical concept through a curriculum series (e.g., How is the concept of *variable* developed in the MiC curriculum?). For a particular mathematical topic, compare the mathematical approaches of different CMs to reform recommendations (e.g., NCTM *Standards)* for that topic. Examine a variety of student work on mathematics problems from the CMs to develop hypotheses about students' learning and implications for teaching. Use the teacher's guide of a CM unit to plan and teach a lesson to peers (preservice teachers).

Curriculum	Compare and contrast the ways that two different CM units approach the teaching and learning of the same mathematical topic.
	Identify the philosophies underlying different curriculum projects and relate them to the ways the CMs position students and the teacher in the learning process.
	Watch a video of a classroom in which a particular lesson from CMs is taught, and (with the assistance of the teacher's guide) identify ways in which the teacher adapted the recommendations of the CMs.
	Adapt traditional textbook lessons to develop technology-rich, inquiry-oriented lesson plans and examine the differences and similarities in the range of lessons developed by preservice teachers.

Appendix B
Sample Questions and Responses to a Common Problem

Questions about the Locker Problem and the Changing Positions Problem	**Sample Responses from Two Preservice Elementary Teachers (from written assignments)**
How are these two problems related? What features of the two problems are similar? What features are different?	• The Locker Problem and the Changing Positions Problem were similar in that they dealt with the idea of perfect squares in a critical thinking way. They both used a game format to make it more appealing while at the same time learning the concept. The difference was that in the Changing Positions Problem, the student actually took part in it in a classroom setting. [Annika] • The Changing Positions and Locker problems were related because they both used factors. The students with a card that had a factor that was called by the teacher sat or stood. The students in the school shut or opened the lockers that they were factors of. [Joy]
How did these problems (and/or the curriculum units they are in) engage you in thinking about each of the following mathematical ideas: factors, factor pairs, and square numbers?	• I think that factors were addressed clearly in *Reflections on Numbers* because they approached it with several different methods. They used graphs and factor trees to demonstrate ways of finding the factors. This gives the students options in finding the best way to solve the problem for themselves. The *Prime Time* book simply used word problems but I found that they didn't give methods in finding the answers as well as the other book did. [Annika] • *Prime Time* addressed factors most directly with the factor game. *Reflections on Number* addressed factors most directly using prime factors with factor trees. Both ways, the game and the factor trees, clearly show

	the factors of a number. The game is beneficial because students get to interact with other students to try and win the game. The factor trees are different from the game because students have to find the factors themselves, while in the game the factors are already there and the students just have to circle them. [Joy]
Did you find the context for one of the problems more accessible than the other?	• I found the Changing Positions problem easier to solve because it was one that you could physically partake in. [Annika] • I thought the Changing Positions problem was easier to solve because a classroom of students is less than 1000. Even though you didn't have to go all the way up to 1000 in the Locker Problem, it was easier knowing in the Changing Positions Problem that you could get the pattern with a small number. [Joy]
Why do you think the Changing Positions Problem appears early in *Reflections on Number* but the Locker Problem appears at the end of *Prime Time*? Which placement makes more sense to you?	• I think that the two books placed the problems in different places in the text depending on how they introduced the concept. If I were to teach this, I may use the problem in the beginning to introduce the topic and make it interesting. Then I would follow up with problems and discussion involving the pattern that occurred. [Annika] • I really don't know why an author would choose this order unless the books are intended for different grade levels. I would prefer to end with this type of problem because it requires a lot of skills of pattern solving with factors and multiples. I think the students should have lots of practice with these concepts before trying to solve a problem as tricky as this. [Joy]

Flowers, J. and Rubenstein, R. N.
AMTE Monograph 3
The Work of Mathematics Teacher Educators
©2006, pp. 29-44

3

A Rich Problem and Its Potential for Developing Mathematical Knowledge for Teaching[1]

Judith Flowers
Rheta N. Rubenstein
University of Michigan – Dearborn

What makes a rich mathematical task for teachers? How do worthwhile tasks play out in preservice classrooms that are centered on problem solving, reasoning, and sociomathematical norms? How do robust problems develop mathematical knowledge for teaching, even in domains beyond those of the original problem? This paper explores perspectives on these questions. It builds from a project reforming the number strand in a curriculum for Mathematics for Elementary Teachers. The paper has implications for mathematics methods as well as content courses.

What makes a rich mathematical task for teachers? For students, rich tasks ask solvers to use non-algorithmic thinking, to find relationships among mathematical ideas, to identify and use relevant knowledge, to apply considerable cognitive effort, and to monitor their own progress (Smith & Stein, 1998). Teacher educators have wondered whether these characteristics are the same or if something different is needed for a mathematical task to be valuable for preservice teachers. In this paper, we share some of our thinking and struggles related to this issue.

Our work is part of a project at the University of Michigan-Dearborn where we are revising the Mathematics for Elementary Teachers curriculum to be problem and reasoning based. (See also Flowers, Kline, & Rubenstein, 2003 and Flowers, Krebs, & Rubenstein, 2006). We build on a modest number of problems that have the potential for developing mathematical proficiency (National

[1] This paper is based on work supported by the National Science Foundation under grant DUE 0310829. Any opinions, findings, conclusions, or recommendations expressed herein are those of the authors and do not necessarily reflect the views of the National Science Foundation.

Research Council, 2001) and mathematical knowledge for teaching (Ball, 2003; Ball & Bass, 2000; Ball, Bass, Sleep, & Thames, 2005). Although prospective teachers need to know the mathematics they teach as well as the foundations and extensions of that mathematics, research indicates that their knowledge of mathematics is limited (Ball, 1990; National Research Council, 2001). In particular, U.S. elementary teachers, in contrast to their counterparts in other countries, lack a profound understanding of fundamental mathematics (Ma, 1999) and struggle with processes of abstraction, generalization, and justification. Moreover, the recommendations of the National Council of Teachers of Mathematics (2000) require more of teachers than before. When children are asked, as is recommended, to reason, represent, communicate, and build connections among mathematical ideas, their teachers need a strong mathematical foundation to analyze student responses (Conference Board of the Mathematical Sciences, 2001; Mathematical Sciences Education Board, 2001; National Research Council, 2001).

Our work also evolves from research and thinking related to sociomathematical norms (Ball & Bass, 2000; Carroll, Mumme, & Romagnano, 2005; Hufferd-Ackles, Fuson, & Sherin, 2004; Lampert, 2001; Yackel & Cobb, 1996). These norms expect prospective teachers to build mathematical understandings individually and collectively by taking responsibility for their own personal learning and that of the community. They need to look to the logic of mathematics to decide what is true, not to an outside authority. Consequently, we expect preservice teachers to work on challenging problems and present publicly their ideas or solutions, however tentative, incomplete, misconceived, or inspired. They bring to bear their own prior knowledge, ideas previously developed, definitions, representations, and strategies. Together, as these preservice teachers explain, listen, reason, represent, question, conjecture, generalize, and justify, they build understanding. For example, multiple strategies are more than welcomed; they are essential learning tools. Analyzing multiple strategies makes apparent levels of sophistication and abstraction, which in turn, provides an opportunity for preservice teachers to make connections among representations and move their own thinking to more highly developed levels. Errors are not just sources of disequilibrium; they are opportunities to reconceptualize problems and explore contradictions and solutions or to pursue alternative strategies.

Additionally, our work builds on the attempts of Ball and others (2005) to define and use the implications of "mathematical knowledge for teaching." Ball and Bass (2000) note that there are many tasks of teaching that require specialized knowledge of mathematics beyond what students are learning or what an educated person might know. Such tasks include recognizing the mathematical potential of a task, understanding the trajectory of a task (how it connects to ideas studied earlier and later), selecting numbers strategically, considering the relative merits of different representations, and anticipating students' responses.

In this paper we share an example of a number theory problem we find mathematically rich. We detail how discussions about the problem typically flow within the sociomathematical norms previously described. Then we illustrate how the number theory within the problem helps develop mathematical knowledge for teaching beyond the original domain into work with fractions.

Overview of the Problem

The problem explored in this paper is Factor Pairs for Multiples of 100 (or, more succinctly, Factor Pairs for 100s). It is adapted from *Mathematical Thinking at Grade 5* (Kliman et al., 1998), one of the resources for our curriculum development. The problem culminates a two-week unit on number theory within a first course in Mathematics for Elementary Teachers. This problem provides an opportunity to apply and consolidate much of the preceding number theory work. In the following sections, we detail the posing of the problem, preservice teachers' approaches, the language and reasoning they use, and our perceptions of the ways their learning connects to what has been developed so far.

Reasoning about Factor Pairs for 100

On the surface, generating factor pairs for 100 seems to be a simple task. Typically preservice teachers march themselves through the counting numbers as potential factors and determine which ones have a "partner." Most "know" the pairs and that they may quit at 10×10. In response to the questions, "How do you know you have all the factor pairs?" and "How far do you need to go to test factors?" preservice teachers need to think deeply. For example, why are there no factors between 50 and 25? How is this related to there not being factors between 2 and 4? Out of this discussion comes the recognition that factors come in pairs, that these pairs have a constant product, that as one factor grows the other gets smaller, and that whole numbers are discrete. Preservice teachers realize that when 10 arises with itself as its "partner" they are done; going beyond 10 corresponds to finding a factor whose partner is less than 10 and they have already found those smaller factors. Drawing out this reasoning is not easy, but it makes the future teachers think about the mathematics.

Also the preservice teachers develop generalizable strategies for finding factor pairs, particularly of 100s. Following are samples of their suggestions and our analysis. (Items are numbered only for readability).

Conjectured Strategy 1. *Factor pairs can come from equivalent products.*

> S: Well if you start with 2 × 50, if you [multiply] 2 times 2, it's a 4, then 50 divided by 2...so 4 times...25 is the partner...If you [make] twice as many groups, you're going to have half [as much] in each group.

In this strategy, preservice teachers use concepts they have explored in earlier work investigating equivalent products arising from multiplying and dividing two factors by the same number.

Conjectured Strategy 2. *Prime factorization may be used.*

> S: I did like 2 × 2 × 5 × 5. That's what I did first....I did the 2 by itself and then ... in parentheses, the 2 × 5 × 5 which would give me 50. ...Then I took the 5 out and then I did 2 × 2 × 5 in parentheses and that gave me...5 × 20. I did the 2 × 5 and then 2 × 5 again, so that gave me 10 × 10 and then the normal 1 × 100. [The student had earlier identified 4 × 25 and had also asserted that 1 was not a prime number.]

The preservice teachers begin to see the structure of numbers through prime factorizations, which they use as a tool to generate factor pairs.

Analyzing the work of the preservice teachers continues to make us think about their progress in many domains. We think of their strategies as *conjectures*; a particular process works and can be justified, so we press them for the reasoning behind the process. They appear to rely on an equal groups meaning of multiplication (rather than area of a rectangle). We aim to have them make connections among the strategies, e.g., seeing how a prime factorization representation could reveal the other strategies. We expect that these insights will develop as work on the problem continues.

Reasoning about Factor Pairs for Hundreds from 200 to 600

Now we set the preservice teachers to work finding factor pairs for the multiples 200 to 600. We make clear that they are not to start from scratch, but are to use the list they have of factor pairs for 100. In effect, we require them to abandon procedures they may have created and use relationships among the pairs. Our intent, as this reframing of the task suggests, is not merely that they develop strategies to find all the pairs; the larger goal is that they find and reason about relationships between factors and multiples. [Appendix A illustrates a solution and is usually provided to preservice teachers subsequent to the entire activity.] Again, we present typical strategies and our thoughts about them.

Conjectured Strategy 3. *Double the second factor of each pair for 100 to get the factor pairs for 200.*

This is the most common strategy that arises for factor pairs for 200. Preservice teachers often develop this conjecture because the factor pairs for 200 start with "1 × 200," which looks like the "1 × 100" in the previous list. Some students connect this relationship to meanings for multiplication that they have developed earlier.

> S: Because 200 is double 100.... You can think of 1 group of 100 things in it, so to get to 200 you double 100 [and] you have to double the number of things in the group. So now you have 1 group of 200.

We are heartened that students realize that using a meaning of an operation (e.g., equal groups) contributes to a good justification.

Other students often suggest doubling the first factor in the pairs for 100. Still others have claimed, "It doesn't matter; you could double either the first factor or the second." At this point, someone usually notes that by doubling just the "second" factors of 100, 8 is missing from the factor pairs. So the conversation continues as students modify their conjectures.

Conjectured Strategy 4. *Double each factor in the pairs for 100 to get the factor pairs for 200. Then remove duplicates.*

Students are usually surprised that simply doubling one factor from their list for 100s would not produce all the pairs. Again, discussion about the roles of both factors in a factor pair or multiplication problem often arises (the number of groups vs. the size of a group), a vital distinction for teachers often blurred in U.S. curricula.

The strategies noted so far involved reasoning within one multiple of 100. Students also see patterns across the hundreds.

Conjectured Strategy 5. *Use patterns with addition.*

> S: So [for factor pairs for 300] 2 groups of 50 things, plus 2 groups of 100 things gives 2 groups of 150 things [e.g., $2 \times 50 + 2 \times 100 = 2 \times 150$].

Here we have the distributive property used as a way to build factor pairs for a multiple of 100 from pairs for previous multiples. Later, in related homework when students study Figure 1, a chart of factor pairs, they continue to think about similar cases.

Analyzing Strategies for Finding Factor Pairs for 600

By the time students work on pairs for 600, one idea they have is to take the "new" factor, and multiply it by each of the factors in every pair from a previous list as they did in conjectured strategy 4. They have seen this work for 200 through 500. Now we ask, "If you are trying to find all the factor pairs of 600, which one list of factor pairs is most helpful, the list for 100, 200, 300, 400, or 500?"

> S1: 200.
> S2: 300 is better than 200 because it has a longer list to start with.
> S1: 200 has fewer pairs, so there's fewer duplicates, so 200 is better.

These comments focus more on efficiency, an attribute of procedures, but less on the logic that might generalize. They illustrate students' understanding that using the factor pairs for 200 misses a factor of 3. Another student's comment moves the discussion toward a general plan.

> S3: I think you can generate the [list] from the 100s. Multiply by 6 on either side.

At this point other students noted that this strategy misses 3×200, 8×75, and 15×40. Then S1 noted that $6 = 2 \times 3$ and you could use 4×25 and produce $(2 \times 4) \times (3 \times 25)$ to get 8×75.

We would like the preservice teachers to realize that 6 being composite makes a difference. We hold this discussion in abeyance and continue with 700. At that point, students usually realize that they have a strategy for 200, 300, 500, 700, 1100. That is, when the "new" factor (the number of hundreds) is prime, they could use the list of 100s and "multiply on either side" by the missing factor. When we ask, "Why is 600 different?" they recognize that 6 being composite impacts the situation. When the preservice teachers examine the prime factorization for 600 they realize, as S1 did, that there are two new prime factors and they might have been split within a factor pair.

One major outcome of this problem is that future teachers appreciate the differences made by primes and composites in some mathematical situations. Another is to understand relationships between factors and multiples. We hope that encounters like these help our students attend to nuances as they engage in the mathematical tasks of teaching.

Analysis of the Problem

We expect the reader senses the potential of the Factor Pairs for 100s problem to bring to the surface many important mathematical concepts and processes and to involve preservice teachers in conjecturing, reasoning, and justification. Appendix B summarizes some of this potential. Although the figure appears as a list, we envision a network of inter-related ideas and dispositions. One of our overarching goals is for our students to build for themselves such "knowledge packages" (Ma, 1999). We continue to ponder ways to help preservice teachers develop these mathematically robust networks.

Surely, we want them to know the mathematics. Elementary teachers must be fluent in the concepts of multiplication, division, factors, multiples, primes, composites, proportionality, and other aspects of mathematical reasoning that this problem embodies. Also, working on an investigation that they might, in fact, do with students (given that the problem is drawn from an existing elementary program) enhances their ability to orchestrate the same task with children. More important, a major part of our philosophy is that reasoning in general, and justification in particular, is the "glue" that helps students build and use mathematical relationships, construct new pathways for themselves, transfer mathematical concepts, and be flexible in mathematical thinking. We also believe that work of this nature helps preservice teachers begin to develop knowledge of how mathematical ideas are related (Ma, 1999) and start to see how reasoning from multiplication plays an important role across a spectrum of mathematical domains.

Developing Knowledge for Teaching

Teachers' preparation needs to provide a great deal more than knowledge of mathematics, even with an emphasis on reasoning. Our goals are broader – we want to prepare teachers for more nuanced work. For example, we want them to be ready to respond to an array of student ideas, to pose mathematically challenging questions (often on the spot), to be sensitive to the choice of numbers in a problem, to be cautious in deciding which representations they might highlight, and to have a sufficiently well-developed and inter-connected understanding of mathematics so that they can use that understanding flexibly and wisely throughout a multitude of teaching tasks. Consequently, we have been challenging ourselves to think deeply about how to achieve these goals. In the specific case of the problem discussed here, we ask ourselves, "How might teachers' work on a rich problem such as Factor Pairs for 100s support their development of the knowledge they need for teaching mathematics?"

Our search has brought us to study more ideas about teachers' mathematical understanding. Hill and Ball (2004) have speculated that how teachers hold knowledge may matter more than how much knowledge they hold. Other researchers (Ball, 1990; Ball & Bass, 2000; Ball, Lubienski, & Mewborn, 2001; Ma, 1999) have suggested that the germane qualities of teachers' knowledge include attributes such as the following: (1) whether it is procedural or conceptual, (2) whether it is connected to big ideas or isolated, and (3) whether it is abstract and compact or conceptually unpacked and meaningful. Consider the Factor Pairs for 100s problem in light of these attributes. Regarding the procedural vs. conceptual attribute, preservice teachers typically begin by using a procedure. However, when we ask them not to use the same strategy for other multiples, they look for and justify relationships, creating a conceptual task. On the second attribute, connected versus isolated, we see the problem as highly connected to many foundational ideas of the elementary school mathematics curriculum: meanings of factors and multiples; meanings of multiplication and division; divisibility properties; the commutative, associative, and distributive properties; prime factorization; and proportional reasoning. On the third attribute, abstract versus unpacked, we see the requirement to provide reasons and justifications, to find relationships and generalizations, as central to the "unpacking" of the mathematics.

Building on the work of Ball (2003), Ball and others (2005), and Ball and Bass (2000) on "mathematical knowledge for teaching," we have developed in Appendix C samples of mathematical tasks for teaching. Our belief is that teachers need support in transforming their understanding of mathematics per se to mathematics as a resource for teaching. Because our students have been developing a reasoning-based background, we believe they may be ready in a methods course to initiate this transformation. In thinking about how to develop such a course, we have been trying to determine how to support the development of mathematical knowledge as a resource for teaching.

Transforming Mathematical Knowledge to a Teaching Resource

How might investigating and reasoning about Factor Pairs for 100s and, more generally, the intense work in number theory that it embodies, help to transfer and transform mathematical knowledge to a teaching resource? We could pose this question by looking for connections in areas like number theory itself, whole number multiplication, or division. Instead, to challenge ourselves and to look for transferability, we explore how number theory knowledge may become a resource in a different domain, fractions. We speculate that this analysis may be helpful in improving both our content course and our methods course. We ask, "What are some teaching problems that might arise in instructional programs for fractions and how does a teacher's number theory knowledge come into play?" Using our analysis from Appendix C, we share a few samples of our response to this question. Comments in *italics* show correlations to tasks identified in Appendix C.

Models for Fraction Addition/Subtraction

Elementary students are beginning to work on adding and subtracting fractions. Their teacher wants to provide them a model that highlights the mathematical underpinnings for a standard algorithm. She considers models in curriculum materials; one suggests an unmarked circle and another suggests a dial clock model. She considers the advantages of each. The unmarked circle has the potential for drawing any possible number of equal divisions, but the teacher knows students struggle to draw congruent portions. The dial clock offers potential denominators of 12 and 60 occurring visually. Recognizing that 12 and 60 are rich in factors, she reasons that students can find common equivalences for halves, thirds, fourths, sixths, twelfths and, with 60ths, even fifths, tenths, twentieths, and others. Sums like $\frac{1}{3}+\frac{1}{6}=\frac{1}{2}$ are natural and develop students' fluency. Sums like $\frac{1}{4}+\frac{1}{3}=\frac{7}{12}$ help develop understanding that addition requires adding equal-sized pieces, a bridge to an algorithm using common denominators. *This case requires the teacher to make selections from curriculum materials, to select numbers judiciously, and to consider the relative merits of different representations.*

The Distributive Property for Fraction Multiplication

A teacher wants to help students be flexible in multiplying mixed numbers by whole numbers. He/she knows that many students think they must first convert the mixed number to an improper fraction. He imagines this misconception is due to their fragile or incomplete understanding of multiplication, and hence, of the distributive property. He designs a calculation for them to do and requires that they use more than one strategy. One strategy he encourages is the use of mental math. He decides to use $3\frac{1}{12} \times 300$, reasoning that 3 and 4 are both factors of 300, so 12 must be a factor too. The teacher believes this problem will reward insightful use of the distributive property. *In this situation, the teacher anticipates students' misconceptions, analyzes their mathematical source, considers ways to address those misconceptions, and judiciously selects numbers.*

Fraction Multiplication Algorithm

A teacher wants to develop the reasoning underlying a standard algorithm for multiplication of fractions using a rectangular model. His experience with primes and composites reminds him that choosing the right numbers matters. He knows that using a pair of fractions with four different primes as numerators and denominators, e.g., $\frac{2}{3} \times \frac{5}{7}$, as opposed to numbers with common factors, e.g., $\frac{2}{3} \times \frac{3}{4}$, will more clearly illustrate the algorithm and has the potential to help students see how the product's numerator and denominator arise. Figure 1 shows the kinds of representations the two problems could produce. *In this case the teacher is judiciously selecting numbers and considering the relative merits of different representations.*

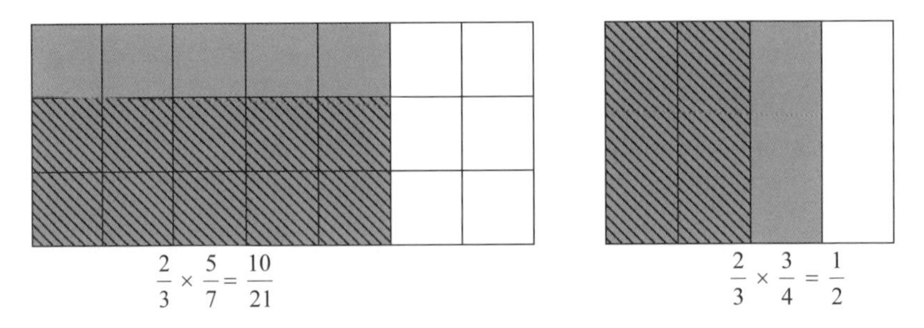

Figure 1. Selecting Numbers for Fraction Multiplication Algorithm

The previous examples stem from our own practice as we plan and teach. Monitoring our own work provides ideas for tasks to use in a reformed methods course that may help preservice teachers transform mathematical knowledge into a teaching resource.

Closing Perspectives

In this article we have illustrated the power for future teachers of exploring a rich problem and clarified the qualities of such problems – the connections to central and foundational mathematical ideas, the potential for a variety of approaches, the opportunities for justification, and the value for the work of teaching. Further, we conjecture that preservice teachers' intense work with number theory has potential to impact mathematical tasks they need to perform in the practice of teaching, even in another strand like fractions. We hope this work will help all of us as teacher educators think carefully about what mathematical knowledge the tasks of teaching entail and to use these insights to plan instruction in both content and methods courses. Overall, we believe there is important potential in a well-developed sequence of content and methods courses, focused on mathematical problem solving and reasoning, situated within an environment of sociomathematical norms, and aimed at developing mathematics as a teaching resource.

References

Ball, D. L. (1990). Prospective elementary and secondary teachers' understanding of division. *Journal for Research in Mathematics Education, 21*, 132-144.

Ball, D. L. (2003). Using content knowledge in teaching: What do teachers have to do, and therefore have to learn? Keynote presented at the Third Annual Conference on Sustainability of Systemic Reform. http://sustainability2003.terc.edu/go.cfm/keynote (as of June 14, 2005).

Ball, D. L., & Bass, H. (2000). Interweaving content and pedagogy in teaching and learning to teach: Knowing and using mathematics. In J. Boaler (Ed.), *Multiple perspectives on the teaching and learning of mathematics* (pp. 83-103). Greenwich, CT: JAI/Ablex.

Ball, D. L., Bass, H., Sleep, L., & Thames, M. (2005). A theory of mathematical knowledge for teaching. Presentation at 15th ICMI Study, *The Professional Education and Development of Teachers of Mathematics*. Sao Paulo, Brazil.

Ball, D. L., Lubienski, S., & Mewborn, D. S. (2001). Research on teaching mathematics: The unsolved problem of teachers' mathematical knowledge. In V. Richardson (Ed.), *Handbook of research on teaching* (Fourth Edition) (pp. 433-456). New York: Macmillan.

Carroll, C., Mumme, J., & Romagnano, L. (2005). The role of sociomathematical norms in mathematics professional development. Presentation at Annual Conference of the Association of Mathematics Teacher Educators, Dallas, TX.

Conference Board of the Mathematical Sciences. (2001). *Mathematical education of teachers.* Providence, RI: American Mathematical Society.

Flowers, J., Kline, K., & Rubenstein, R. (2003). Developing teachers' computational fluency: Examples in subtraction. *Teaching Children Mathematics, 9*(6), 330-334.

Flowers, J., Krebs, A., & Rubenstein, R. (2006). Problems to deepen teachers' mathematical understanding: Examples in multiplication. *Teaching Children Mathematics*, *12*(9), 478-484.

Hill, H. C., & Ball, D. L. (2004). Learning mathematics for teaching: Results from California's mathematics professional development institutes. *Journal for Research in Mathematics Education, 35* (5), 330-351.

Hufferd-Ackles, K., Fuson, K. C., & Sherin, M. G. (2004). Describing levels of components of a math-talk community. *Journal for Research in Mathematics Education, 35* (2), 81-116.

Kliman, M., Tierney, C., Russell, S. J., Murray, M., & Akers, J. (1998). *Mathematical thinking at grade 5.* White Plains, NY: Dale Seymour Publications.

Lampert, M. (2001). *Teaching problems and the problems of teaching.* New Haven, CT: Yale University Press.

Ma, L. (1999). *Knowing and learning mathematics for teaching: Teachers' understanding of fundamental mathematics in China and the United States.* Mahwah, NJ: Lawrence Erlbaum Associates.

Mathematical Sciences Education Board. (2001). *Knowing and learning mathematics for teaching.* Washington, DC: National Academy Press.

National Council of Teachers of Mathematics. (2000). *Principles and standards for school mathematics.* Reston, VA: Author.

National Research Council. (2001). *Adding it up: Helping children learn mathematics.* J. Kilpatrick, J. Swafford, B. Findell (Eds). Mathematics Learning Study Committee, Center for Education, Division of Behavioral and Social Sciences and Education. Washington, DC: National Academy Press.

Smith, M. S., & Stein, M. K. (1998). Selecting and creating mathematical tasks: From research to practice. *Mathematics Teaching in the Middle School, 3*(5), 344-350.

Yackel, E., & Cobb, P. (1996). Sociomathematical norms, argumentation, and autonomy in mathematics. *Journal for Research in Mathematics Education*, *27*, 458-477.

Judith Flowers teaches mathematics education courses at the University of Michigan – Dearborn. She has had experience in school level teaching, curriculum development, preservice teacher education, and professional development. She has researched relationships among children's understanding of multiplication, division and proportional reasoning. Recently she has been researching preservice teachers' developing mathematical justification. Currently, she is co-directing, with Rheta Rubenstein, a NSF-supported curriculum development project, *A Reasoning and Problem Solving-based Curriculum for Elementary Mathematics Teachers*.

Rheta N. Rubenstein is professor of mathematics at the University of Michigan-Dearborn where she collaborates with Judith Flowers on curriculum development for the mathematics for elementary teachers program. She also coordinates a graduate program with a specialty in middle grades mathematics. Formerly she taught at Schoolcraft College, University of Windsor, and the Detroit Public Schools. She was editor of the NCTM 66th (2004) yearbook, *Perspectives on the Teaching of Mathematics,* and is currently general yearbook editor for NCTM's 2008-2010 yearbooks. She is co-author of several high school textbooks and a mathematics methods textbook for middle grades teachers.

Appendix A
Factor Pairs of Multiples of 100

100	200	300	400	500	600	700	800	900	1000
1 × 100	1 × 200	1 × 300	1 × 400	1 × 500	1 × 600	1 × 700	1 × 800	1 × 900	1 × 1000
2 × 50	2 × 100	2 × 150	2 × 200	2 × 250	2 × 300	2 × 350	2 × 400	2 × 450	2 × 500
		3 × 100			3 × 200			3 × 300	
4 × 25	4 × 50	4 × 75	4 × 100	4 × 125	4 × 150	4 × 175	4 × 200	4 × 225	4 × 250
5 × 20	5 × 40	5 × 60	5 × 80	5 × 100	5 × 120	5 × 140	5 × 160	5 × 180	5 × 200
		6 × 50			6 × 100			6 × 150	
						7 × 100			
	8 × 25		8 × 50		8 × 75		8 × 100		8 × 125
								9 × 100	
10 × 10	10 × 20	10 × 30	10 × 40	10 × 50	10 × 60	10 × 70	10 × 80	10 × 90	10 × 100
		12 × 25			12 × 50			12 × 75	
						14 × 50			
		15 × 20			15 × 40			15 × 60	
			16 × 25				16 × 50		
								18 × 50	
			20 × 20	20 × 25	20 × 30	20 × 35	20 × 40	20 × 45	20 × 50
					24 × 25				
						25 × 28	25 × 32	25 × 36	25 × 40
								30 × 30	

Appendix B
Sampling of the Mathematics within the Factor Pairs for 100s Problem (Using the Proficiency Strands from *Adding It Up*, National Research Council, 2001)

Conceptual Understanding – *comprehension of mathematical concepts, operations, and relations*

Multiplication, meanings - equal groups or areas of rectangle
Equivalent products
Division, meanings – partitive, quotitive
Divisibility tests
Remainders and how they can be used to predict
Links between multiplication and division
Factors – deepens meanings
Multiples– deepens meanings
Relationships between factors and multiples
Direct proportional relationship for factors across lists
Inverse proportional relationships of factors within a list
Prime factorization
Properties – commutative, associative, distributive

Procedural Fluency – *skill in carrying out procedures flexibly, accurately, efficiently, and appropriately*

Fact fluency
Familiarity with factors of multiples of 100
Mental math

Strategic Competence – *ability to formulate, represent, and solve mathematical problems*

Make the problem simpler
Use a related problem
Search systematically
Solve a second way to double check
Try to see the problem in a different way (e.g., prime factorization)
Look back; try to see more once a problem is solved

Adaptive Reasoning – *capacity for logical thought, reflection, explanation, and justification*

Potential for conjectures, generalizations, and justifications

Productive Disposition – *habitual inclination to see mathematics as sensible, useful, and worthwhile, coupled with a belief in diligence and one's efficacy*

Persistence
Curiosity
Self-questioning
Self-monitoring

Appendix C
Selected Teacher Tasks Requiring Mathematical Knowledge for Teaching

Tasks related to using instructional materials

More Broad

- Identifying the mathematical potential of a task or problem
- Making selections from curriculum materials
- Understanding what a task develops from and where it leads (curriculum sequence)

Less Broad

- Choosing numbers judiciously when posing a problem or a counterexample
- Understanding the relative difficulty level of related mathematical tasks
- Considering the relative merits of different representations of mathematical ideas
- Appreciating the careful selection of numbers within some curriculum materials

Tasks related to students' mathematical thinking

- Anticipating typical responses to specific tasks
- Anticipating common misconceptions, analyzing their mathematical source, and considering ways to address those misconceptions
- Anticipating creative solution strategies

Tasks related to bridging students' thinking and mathematics

- Being prepared to support students' responses at a variety of levels
- Knowing a spectrum of mathematical language (from less to more formal) that might arise in the midst of a particular task
- Knowing the relative merits of different representations of mathematical ideas
- Listening to students and figuring out what they are thinking
- Navigating a productive discussion
- Creating a problem mathematically similar to another but slightly easier or more challenging
- Deciding among alternative courses of action
- Being able to explain a mathematical idea

Loe, M. and Rezac, L.
AMTE Monograph 3
The Work of Mathematics Teacher Educators
©2006, pp. 45-61

4

Creating and Implementing a Capstone Course for Future Secondary Mathematics Teachers

Melissa Loe
Lisa Rezac
University of St. Thomas

The Conference Board of the Mathematical Sciences [CBMS] recommends a capstone experience for future teachers of secondary mathematics. This paper describes the development and implementation of such a capstone course for undergraduates which involves study, discussion, and presentation of mathematics topics directly related to secondary mathematics as recommended by the CBMS. The course also meets licensure requirements set forth by the Minnesota Board of Teaching. We discuss standards and recommendations which influenced the creation of the course, as well as course goals and components. We also share our successes, failures, preservice teachers' reactions, and proposed changes in the course.

In the first AMTE monograph, Mathews (2004) describes the development of a *Concepts in Calculus Course for Middle School Teachers* in response to the recommendations from both the Mathematical Association of America [MAA] and the Conference Board of the Mathematical Sciences [CBMS]. At the University of St. Thomas, we have developed and implemented a course to respond to other recommendations in the 2001 CBMS document, *The Mathematical Education of Teachers* [MET]. In our course, we aim to follow the recommendations of the CBMS and address the Minnesota Board of Teaching [MNBoT] licensure requirements for future high school teachers.

Part I of the MET document contains two main recommendations for the preparation of future high school teachers. The first focuses on redesigning core mathematics courses to address connections between the college course content preservice teachers learn and the high school course content they will teach. The second recommendation suggests that mathematics departments "support the design, development, and offering of a capstone course sequence for teachers in which conceptual difficulties, fundamental ideas and techniques of high school mathematics are examined from an advanced standpoint" (CBMS 2001, p. 39).

To satisfy the MNBoT licensure requirements, which are content driven and not course prescribed, in 2000 we required our secondary mathematics education majors to take 12 four-credit semester courses. These courses consist of three semesters of calculus and one semester each of linear algebra with differential equations, abstract algebra, real analysis, probability, statistics, geometry, computer programming, advanced discrete mathematics, and a capstone course. By 2002, we realized that we did not have the enrollment necessary to continue the advanced discrete mathematics course, so we moved the required discrete topics to the capstone course. Although this solved our enrollment issues, the addition of the discrete topics led to an overloaded capstone course, especially when we believed two semesters were needed for a capstone course that fully meets MET document expectations.

There is at least one textbook that satisfies the MET document recommendations (Usiskin, Peressini, Marchisotto & Stanley, 2003). However, in order to meet the additional requirements of the MNBoT, we realized we needed to design our own course.

Specifics of the Capstone Course

Timing, Duration and Frequency

Currently, our capstone course, Math 450 - *Advanced Mathematics: Exploration and Exposition*, is a four-credit course, taught in the department of mathematics. It is offered one semester each year and is required of mathematics majors who earn a secondary education co-major. Preservice teachers typically take the course during the semester preceding their student teaching. The class meets for 65 minutes three times a week for 15 weeks and usually attracts 8-12 secondary mathematics education majors. The capstone course has been offered four times from 2000 to 2004.

Goals for the Capstone Course

There are four main goals for the course:

1. To provide bridges among upper-level mathematics courses, especially real analysis, abstract algebra, and geometry.

2. To provide preservice teachers an opportunity to explore connections to the high school curriculum so that they have a better understanding of the mathematics they will teach.

3. To provide preservice teachers with additional exposure to areas of mathematics in which they may be deficient, according to the current course requirements at our university and our interpretation of the National Council of Teachers of Mathematics [NCTM] and MNBoT standards (i.e., discrete mathematics and the history of mathematics).

4. To provide preservice teachers experiences with research and writing in mathematics and oral presentations to their peers and instructors.

We assess the extent to which our course goals are met through two major individual projects, homework assignments, class participation, occasional quizzes, class discussions, and a final exam.

Capstone Content Components

The course has several overlapping components: bridging content areas of mathematics, assessment topics and ties to the high school curriculum, exposition - historical research, exploration - contemporary research, discrete mathematics, and topics in "higher mathematics." We describe each of these components in some detail.

Building Bridges via Fermat's Last Theorem. By developing the theme, "Bridges through Number Theory," we expose our preservice teachers to a study of the history of mathematics, and bring that history alive by reading and discussing Singh's book, *Fermat's Enigma* (1997). The subject of the book, Fermat's Last Theorem, is an easily stated number theory problem, yet the proof required bridging several areas of mathematics, including algebra of the real number system, abstract algebra, and real analysis. We believe that several of the goals of the course are reached by examining these bridges in the historical context the book provides. In addition, there are connections to the high school curriculum through modular arithmetic, solutions to algebraic equations, the Pythagorean Theorem and other theorems in geometry, and symmetry in various areas of mathematics. The preservice teachers read *Fermat's Enigma* over the course of the semester, and prepare answers to discussion questions, ranging from an explanation of the Fundamental Theorem of Algebra to a description of the continuum hypothesis. They lead discussions at several class meetings throughout the term. About half of the discussion questions are elementary, to ensure that the preservice teachers catch certain facts from the reading, and the other half are questions which can lead to extended searches outside of the book. (See Appendix A for selected examples of the second type.)

By tracing the development of the proof of Fermat's Last Theorem, Singh blends history with current unsolved problems and well-known mathematics from hundreds or thousands of years ago. By reading some of Andrew Wiles' own words describing his struggles with the theorem, our preservice teachers come to understand the sacrifice and excitement that some mathematicians experience. Their reactions to the book and discussions are quite positive: many cite "reading Fermat" as a valuable course experience.

By discussing the struggles that Wiles experienced and his devotion to the problem, our preservice teachers gain an appreciation for *doing* mathematics. Singh details the rich history of the problem, including the establishment of the Wolfskehl prize, anecdotes of false proofs, and contributions of numerous mathematicians. Reading the book helps our preservice teachers realize that mathematics is not a solitary pursuit. While reading the book, each keeps a "timeline" of mathematics and mathematicians that is graded for completeness at the last discussion period. Other books that could be used to achieve the same goals are listed in Appendix B.

Analysis of Teaching / Learning Cases and Ties to the High School Classroom. Making connections to the high school curriculum and classroom is a primary goal for the course and accomplished by using the book, *Windows on Teaching Math: Cases of Middle and Secondary Classrooms* (Merseth 2003). Merseth offers eleven case studies on issues in the teaching and learning of mathematics accompanied by a set of discussion questions and a facilitator's guide. The case studies offer a variety of topics so individuals designing their own capstone courses will be able to choose case studies to meet their own goals. Our preservice teachers read the case studies, work through the mathematics involved, prepare answers to discussion questions, and are often responsible for leading discussions. [Another source of case studies is Friedberg's *Teaching Mathematics in Colleges and Universities* (2001). Friedberg's cases were developed to train teaching assistants at the post-secondary level, but are readily adaptable for high school connections because they discuss issues related to teaching algebra and calculus.]

By reading and discussing case studies, our preservice teachers become aware of various assessments they will need to use in the classroom. Although *assessment* often refers narrowly to the evaluation of students' learning, educators also need to assess classroom culture, their own performance, and curricular material. It is this wider interpretation of assessment that we address in the capstone course through case studies. We attempt to bring a future "real-life" situation into our classroom and discuss how we view the problem and how we might address it. Discussion and evaluation of different effective solutions will, in theory, enable better responses than being unprepared, making snap judgments, and missing "teachable moments" in the future classroom.

We also want our preservice teachers to assess textbooks and curricular materials. Accordingly, during one class activity we examine different calculus books and help our preservice teachers trace definitions and theorems through each text to consider the different approaches. For example, we compare and contrast definitions of limit and continuity as well as the Intermediate Value Theorem and the Mean Value Theorem. We ask questions such as: "What level of rigor is applied?" "Are proofs supplied, and if so, in what order?" "Are they accessible to a high school student?" Through this activity, we highlight the differences between traditional and reformed texts, and review critical advanced mathematics that our preservice teachers may not have yet mastered.

One final assessment activity is the creation of scoring guides for exploratory and expository projects. In the 2001 capstone course, preservice teachers helped to create the scoring guides after seeing the project assignments. They first created and shared individual rough drafts and then worked together on final versions to be used by their peers and the instructor for the oral presentation, and by the instructor for the written papers. We now distribute the scoring guides (see Appendix E for an example) with the project assignments so preservice teachers know how they will be assessed. The written peer evaluations of oral presentations are typed and provide anonymous feedback for the presenters.

Expository Project. The first of two major paper/presentation assignments requires each preservice teacher to write a paper and give a class presentation on a mathematics topic that is relevant to the high school curriculum. We adopt the guidelines for authors from the Editorial Policy in *Mathematics Magazine* (n.d.) as our ideal (these can be found in the first paragraph on the inside cover of each issue), replacing "undergraduate" with "high school."

Copies of the *Minnesota K-12 Mathematics Framework* (Minnesota Department of Education, 1998) are available to the preservice teachers as they make choices about topics and prepare their papers and presentations. Our guidelines make clear that these 30-minute expository talks are to be interesting and informative, and must include an activity. Suggestions for activities include diagrams, details of some proof for the class to complete, a worksheet, or student-work at the white board. We urge presenters to use manipulatives to engage the class in the work. We also suggest that an outline helps focus class attention on the important points of the presentation. Preservice teachers are required to use at least three sources for the project and include a complete bibliography. We caution against internet sites as primary resources as these can be unreliable. Often, our preservice teachers who are in the midst of their clinical or student teaching experiences use their consulting classroom teachers as additional resources. The expository project assignment can be found in Appendix D.

Each student meets with the instructor within the first two weeks of the project assignment to prevent duplication of topics and to discourage procrastination. An additional meeting is scheduled one to two weeks before the presentation date for "final adjustments," implying that the work is to be nearly done by this point in time. We urge preservice teachers to consult with us throughout the process of preparing the paper and presentation.

Exploratory Project. The second project assignment requires reading contemporary research in mathematics and explaining that research to an audience of peers. To begin, the instructor brings dozens of current issues of mathematics journals (e.g., *The College Mathematics Journal*, *The American Mathematical Monthly*, *Mathematics Magazine*) to class. Each preservice teacher spends half an hour looking through journals and chooses three articles to read, followed by a report back to the class on the contents of his/her articles the next week. These reports are brief, and are intended to offer an alternative research article to a preservice teacher who might not be satisfied with his/her existing choice.

Preservice teachers then study and digest their articles. Over the next four weeks, they meet individually several times with the instructor to summarize important points, to get clarification and assistance filling in missing details, and to plan the presentation. In the seminar-style 30-minute presentation, the presenter summarizes and explains the research, and addresses possible uses or adaptations of topics from his/her article to a high school classroom. In Appendix C we share titles of some research articles preservice teachers have presented and in Appendix E we provide a copy of the scoring guide used to evaluate their presentations.

Discrete Mathematics. In the capstone course, our preservice teachers work with topics in graph theory and general counting methods for about a third of the semester. Additionally, at least one of the two major papers deals with a discrete mathematics topic. We have used the first five chapters of Tucker's *Applied Combinatorics* (2001) for the course work, and have taken some project ideas from topics in later chapters.

Other Topics in Higher Mathematics. During one offering of the capstone course, preservice teachers did extended work with induction and recursion, and studied the topic of divisibility. Although they had worked with these topics in other advanced mathematics classes, we wanted them to have a deeper and broader exposure. We used the treatment in Usiskin et al. (2002) to look at multiple solutions to problems and to analyze these solutions from different mathematical perspectives. This analysis helped achieve our goal of building bridges among college level courses and the high school curriculum.

Observations and Proposed Improvements

We regard our capstone course as a "work in progress," incorporating changes each time it is taught by considering preservice teachers' reactions and our own evaluations to improve the course. In this section, we share our successes, failures, reactions, and proposed changes. Preservice teachers' papers and presentations show significant improvement from the first project to the second. Requiring the second project allows the use of classmates' and instructor's feedback on the first project, and their own evaluations of others' work to create better, stronger projects. Additionally, some preservice teachers, whose performance in other mathematics courses has been only average, excel on the research projects and presentations. Work on these projects is exciting and rewarding for both the preservice teachers and the instructor. The success that our preservice teachers have with their research projects helps them realize that they can understand current mathematics well enough to present and write about it. Many of the journal articles require a synthesis of information and methods that preservice teachers have used in other mathematics classes. They make the connections with their mathematics classes and fill in gaps in the papers themselves. They really *do the math.*

Another benefit of participation in the capstone course is related to future professional development. When the capstone course was taught in fall 2004, seven of the eight preservice teachers enrolled in the course volunteered at the regional NCTM conference in Minneapolis. They attended sessions and were exposed to some of the "movers and shakers" in NCTM. The preservice teachers were enthusiastic about their experience and future participation in professional organizations.

In spite of these successes, we have had struggles and disappointments with our capstone course. Because we want preservice teachers to do sophisticated mathematics, this course carries a heavy grading demand on the instructor, unlike any other we have experienced. Additionally, a few preservice teachers each semester delay work on the major projects until a week or so before the due date. This procrastination results in poor quality work, stress, and decreased attention given to other work for the course. We have mandated meetings with the instructor in an attempt to prevent procrastination, but this is not a perfect solution. Perhaps most importantly, we believe that connections to the high school curriculum and classroom need to be more explicit. We have considered involving a "consulting" high school teacher in hopes of making stronger, deeper connections.

We reiterate that a capstone course should be taken after other mathematics content courses, and should build bridges between these content courses as well as filling in gaps from within courses. The need to cover new material (e.g., discrete mathematics) means less attention is given to forging those links. We also continue to struggle with the final exam, which depends heavily on preservice teachers' projects and needs careful attention each time the course is taught.

From a preservice teacher's perspective, difficulties with the course usually are related to time management. We continue to search for better ways to eliminate procrastination by setting and meeting intermediate goals. Further, because the course has multiple components, preservice teachers usually are working on several different assignments at any given time. They cite this as a negative aspect, and sometimes suggest that we should do one thing at a time, with a "rest" between assignments. The challenge for us as instructors is to help our preservice teachers manage multiple assignments concurrently, and to see the course as a tapestry of experiences and learning rather than a collection of independent, disconnected components.

Conclusion

In our capstone course, we hope to prepare preservice teachers to understand the mathematics they will teach. Although we continue to develop and improve the capstone course, we believe our course successfully fulfills the dual role of following CBMS recommendations and MNBoT requirements. Capstone courses at other institutions will not follow this model exactly, but we hope that this article is a useful resource for those departments with courses in the development stage. After four implementations and subsequent revisions of the course, we believe that our capstone course improves the cohesiveness of our secondary mathematics education program and strengthens the preparation of our future teachers.

References

Conference Board of the Mathematical Sciences. (2001). *The mathematical education of teachers.* Providence, RI: American Mathematical Society.

Friedberg, S. (2001). *Teaching Mathematics in colleges and universities: Case studies for today's classroom.* Providence, RI: American Mathematical Society.

Mathematics Magazine. Editorial Policy (inside front cover of each issue). Also see guidelines retrieved May 31, 2004 from http://www.maa.org/pubs/mm-guide.html.

Mathews, S. M. (2004). The experiences in a concepts in calculus course for middle school mathematics teachers. In T. Watanabe and D. R. Thompson (Eds.), *The work of mathematics teacher educators: Exchanging ideas for effective practice (AMTE Monograph 1)* (pp. 67-85). San Diego, CA: Association of Mathematics Teacher Educators.

Merseth, K. K. (2003). *Windows on teaching math: Cases of middle and secondary classrooms.* New York: Teachers College Press.

Minnesota Board of Teaching (n.d.). Minnesota Rules and Statutes, Chapter 8710.4600. Retrieved May 31, 2004, from http://www.revisor.leg.state.mn.us/arule/8710/4600.html

Minnesota Department of Education. (1998) *Minnesota K-12 Mathematics Framework.* St. Paul, MN: Author. Retrieved May 31, 2004 from http://www.scimathmn.org

Singh, S. (1997). *Fermat's enigma: The epic quest to solve the world's greatest mathematical problem.* New York: Anchor Books.

Tucker, A. (2002). *Applied combinatorics.* New York: John Wiley and Sons.

Usiskin, Z., Peressini, A., Marchisotto, E. A., & Stanley, D. (2002). *Mathematics for high school teachers: An advanced perspective.* Princeton, NJ: Prentice Hall.

Lisa Rezac, Associate Professor of Mathematics at the University of St. Thomas in St. Paul, Minnesota received her Ph.D. in mathematics in 2000 from the University of Nebraska – Lincoln. She teaches a variety of undergraduate mathematics courses, including the content courses required for teaching licensure for both elementary and secondary preservice teachers. She has been involved in the PREP/PMET programs offered through the Mathematical Association of America and is also involved in summer workshops for inservice teachers. She directs an annual summer math camp for high school girls and has co-led a mathematics study abroad course in Spain.

Melissa Shepard Loe earned her Ph.D. in mathematics from the University of Minnesota in 1990. She is an Associate Professor of Mathematics at the University of St. Thomas where she enjoys teaching all levels of mathematics. During her tenure at St. Thomas, she has worked with K-12 teachers in summer science and mathematics workshops, and has conducted several summer geometry workshops for middle school teachers. She has served as Vice President of the Minnesota Council of Teachers of Mathematics. In her spare time, she enjoys playing the piano, volleyball, running, skiing and keeping up with her husband and four daughters.

Appendix A
Selected Discussion Questions for *Fermat's Enigma* (Singh, 1997)

- To be collected when we finish the book: A timeline of mathematics and mathematicians discussed in *Fermat's Enigma* (1997). Work on this as we read through the book.
- What is the difference between mathematical proof and scientific proof?
- State Fermat's Last Theorem. How is it different from Pythagoras' theorem? Describe concretely what would exist for the $n = 3$ case if the theorem were not true. Who has proved versions of Fermat's last theorem?
- Prove the Pythagorean Theorem. Find a different proof of the Pythagorean Theorem, and bring it to share. (Be able to explain your proofs at the blackboard.)
- What importance is attached to proof by contradiction? Share at least one other application of this technique that you have seen.
- How might David Hilbert's clerk at the hotel deal with the arrival of:
    a. Two coaches, each with an infinite number of guests?
    b. n coaches, each with an infinite number of guests?
    c. An infinite number of coaches, each with an infinite number of guests?
- What is the Fundamental Theorem of Arithmetic? What is the Fundamental Theorem of Algebra? What is the Fundamental Theorem of Calculus? Which of these are discussed in high school? Which of these have you proved? Look up proofs and be ready to discuss them. (Are they at a level you understand? What more would you need to know?)
- Describe the effects of World War II on the public perception of mathematics.

Appendix B
Books with a Historical Perspective on Mathematics

Du Sautoy, M. (2003). *The music of the primes: Searching to solve the greatest mystery in mathematics*. New York: HarperCollins Publishers.

Joseph, G. G. (2000). *The crest of the peacock*. Princeton, NJ: Princeton University Press.

Singh, S. (2000). *The code book: The science of secrecy from ancient Egypt to quantum cryptography*. New York: Anchor Books.

Wilson, R. (2004). *Four colors suffice: How the map problem was solved*. Princeton, NJ: Princeton University Press.

Appendix C
Selected Exploratory Project Research Articles

Agard, D., & Shackleford, M. (2002). A new look at the probabilities in bingo. *College Mathematics Journal*, *33*, 301-305.

Apostol, T., & Mnatsakanian, M. (2002). Surprisingly accurate rational approximations. *Mathematics Magazine, 75*, 307-310.

Bailey, H. (2002). Areas and centroids for triangles within triangles. *Mathematics Magazine*, *75*, 368-372.

Boskoff, W., & Suceava, B. (2004). When is Euler's line parallel to a side of a triangle? *College Mathematics Journal*, *35*, 292-296.

Cairns, G. (2002). Pillow chess. *Mathematics Magazine*, *75*, 173-186.

Redmond, C. (2003). A natural generalization of the win-loss rating system. *Mathematics Magazine, 76*, 119-126.

Appendix D
Expository Project Assignment: Description and Expectations

The purpose of your expository project/talk is to EXPLAIN – in a broad sense – the topic that you have chosen. This is in contrast to a research project, or talk, in which you will present – in a very narrow sense – the content of your research article. The *Mathematics Magazine* is an expository journal (even though we are using some articles from there for the research projects). Their policy is a good start at explaining what exposition entails:

> *Mathematics Magazine* aims to provide lively and appealing mathematical exposition. The Magazine is not a research journal, so the terse style appropriate for such a journal (lemma-theorem-proof-corollary) is not appropriate for the *Magazine*. Articles should include examples, applications, historical background, and illustrations, where appropriate. They should be attractive and accessible to undergraduates and would, ideally, be helpful in supplementing undergraduate courses or in stimulating student investigations. Manuscripts on history are especially welcome, as are those showing relationships among various branches of mathematics and between mathematics and other disciplines. (*Mathematics Magazine*, n.d., Editorial Policy inside front cover)

Think about that description, but replace "undergraduate" with "high-school." For sources on what the Minnesota high school curriculum should entail, you can refer to the *Minnesota K-12 Mathematics Framework* (MN Dept. of Education, 1998). You MUST incorporate some application for high school students into your project.

Your expository talks should be INTERESTING and INFORMATIVE. You should read as much as possible on your topic, then sit back and think – "What was the most interesting thing that I've just learned?" "What was the most important thing that I've just learned?" Then you should focus your project on that idea. Your project will consist of not only presenting the actual idea, but in giving us enough explanation on the background or history so that we can all grasp the idea. Another thing that should be included in your presentation is some interactive activity – perhaps a handout (timeline, diagram, details, worksheet), or a problem you present on the board or overhead for us to work through. This will keep the presentation interesting, and help you to engage your audience. Manipulatives are also a good way to engage the audience. Giving the main points of your talk on an overhead while you elaborate on them is another great way to communicate in a 30 to 40-minute talk.

I've compiled some hints and directions for mathematical talks in another packet. Please read through them to get an idea of what is expected of mathematicians. We will discuss these, so highlight and write down any questions these guidelines raise. In general, talks should have a clear introduction (***make it interesting and attention getting***), a well organized body, and a clear conclusion (which points out exactly what you were trying to explain).

Appendix E
Exploratory Project Presentation Scoring Guide

Introduction **5**

Do you get our attention immediately?
Are your main points previewed?

Body/Organization **20**

Are the ideas presented in a logical order?
Are there smooth transitions?
Does the presentation seem well-planned?

Body/Mathematics **50**

Is there a summary of assumptions?
Is the presenter comfortable with enough background information to understand the topic?
Does (s)he incorporate individual research and knowledge with the article topic?
Is there a clear description of propositions and proposed results?
Are there well-outlined, complete proofs with necessary definitions?
Does the presenter find balance between rigorous proof and sketching the idea of a proof?
Is there a summary of previous work that historically precedes the topic?
Does the presenter understand mathematical justifications and present them clearly?
Are there theorems and proofs presented?
Are applications of theorems presented?
Does the presenter discuss "what tools come together" from our mathematical education?

Conclusion **5**

Are the main points reviewed?
Does the conclusion support the body of the presentation?

Style **10**

Is the presenter well poised and professional?
Does the presenter avoid using "fillers": "umm", "you know", "whatever", "etc.", "Ya"?
Does the presenter make sufficient contact with the audience? (eye or otherwise)

Is (s)he enthused/excited about the topic?
Does (s)he respond well to questions?
How is the tone of the presentation and is it directed at the audience?
Does the presentation go beyond the chalkboard to other forms of media/visual aids?

Activity **10**

Is the mathematics strongly related to the presentation topic?
Does it catch the interest of the audience? Is the activity creative?
Are the students engaged in learning? Is the activity aimed at an appropriate level?
Does the presenter discuss possible adaptations for a secondary level?
Are different learning styles addressed?

Kress, L. and Breyfogle, M. L.
AMTE Monograph 3
The Work of Mathematics Teacher Educators
©2006, pp. 63-78

5

Renewing the Conversation about Gender Equity in Teacher Education

Lauriann Kress
Columbia University

M. Lynn Breyfogle
Bucknell University

In keeping with the National Council of Teachers of Mathematics' Equity Principle (2000) and position statement on Mathematics Education for Underrepresented Groups (1998), this chapter examines gender equity trends in curriculum for prospective teachers. The intent of the chapter is to provide background and raise questions for teacher educators to consider, revisit, and discuss with regard to gender and its place in mathematics teacher preparation. Responses to a survey from members of the Association of Mathematics Teacher Educators are analyzed and summarized in three categories: the need for gender education, the lack of standardization regarding equity issues, and incorporation of gender equity into teacher education curricula. National research, suggested readings, and resources are integrated into each theme.

The National Council of Teachers of Mathematics (NCTM) has provided direction for refining and improving mathematics education since 1920. In terms of equity, NCTM has made clear recommendations, especially in the *Principles and Standards for School Mathematics* (NCTM, 2000) in which equity is included as the first of six principles for the teaching and learning of mathematics. The Equity Principle recommends that teachers provide challenges, opportunities, access to technology, and support for students to learn mathematics. Central to this principle are teachers' high expectations for all students. However, several groups of students, including "students who live in poverty, students who are not native speakers of English, students with disabilities, females, and many nonwhite students" (p. 13), historically have experienced low expectations. These low expectations are often reflected in many students' perceptions of self in terms of mathematical achievement and affect.

This focus on equity was evident before the *Principles and Standards* in the NCTM position statement on "The Mathematics Education of Underrepresented Groups" which identifies several groups as being underrepresented in mathematics-based disciplines: "African-Americans, Hispanics/Latinos, Native Americans, Alaskan Natives, Pacific Islanders, females, children in poor communities, children with disabilities, [and] Asian Americans, among others" (NCTM, 1998). NCTM suggests that educators need to motivate every student to pursue further study of mathematics, examine current policies or programs that potentially lead to math avoidance, diminish mental and institutional barriers that prevent the study of mathematics, and explain to members of underrepresented populations and their families the importance of studying mathematics.

How do mathematics teacher educators unpack what equity means and how can this principle be made alive for prospective teachers? What types of experiences are provided? What readings can be useful in creating a framework of understanding for prospective teachers? This paper focuses only on gender equity in learning to teach secondary school mathematics.

To explore how mathematics teacher educators attend to issues of equity in preparing secondary school teachers, a survey was created (see Appendix A) and sent to approximately 620 members of the Association of Mathematics Teacher Educators (AMTE) and their colleagues. A total of 170 responses (27.4%) were received. However, because the survey focuses on secondary mathematics methods courses, only half of those responses (13.7%) were considered appropriate for analysis. One could argue that the most interested individuals in the topic of gender equity were more likely to respond to the survey. Hence, the results discussed here may be biased toward those situations in which equity *is* addressed; shortcomings in these responses may provide a basis for dialogue about gender equity education in the preparation of prospective mathematics teachers.

Although most survey responses were "yes" or "no", participants had an opportunity to elaborate on their answers through the final open-ended question. An analysis of these responses found similar comments that were grouped into categories: the need for gender education, lack of standardization regarding gender equity issues, and incorporation of gender equity into prospective teacher education curricula. Excerpts from the survey are used to discuss and illustrate these three themes.

Themes within the Survey Responses

Theme #1: Need for Gender Equity

Some respondents reported that gender equity must be included in the preparation of prospective teachers. Others replied just as confidently that gender equity was no longer an important issue to be addressed in mathematics teacher education. Which of these conflicting views is correct? Is there a need for gender equity education? Examining survey arguments may help illuminate thoughts regarding the need for gender equity.

Based on the low response rate and tone of the responses, one interpretation of the results is that teacher educators do not view gender equity as an especially important issue. Of the n = 85 responses, approximately 18% included language suggesting that gender equity was a topic of the past or an unnecessary issue. One respondent stated, "I think gender equity is an important topic, but I don't think it is quite as important as it was 25 years ago. I think we have made great strides in this area" (Survey Response 79 (SR 79)). Other comments focused on an increase in the number of females who are choosing to pursue mathematics. For example, "The majority of our [undergraduate] students [at my institution], who receive majors in both math and math education, are female" (SR 6). Teacher educators who are sympathetic to gender equity issues appear to have seen great strides in gender equity, so that the need for discussion no longer exists.

At the same time, however, none of the respondents viewed gender equity as "not important" and over 40% classified it as "very important." In comments, one respondent stated, "an important topic like gender equity often does not get the attention it needs" (SR 81) and another stated, "I address this in all mathematics, technology for teachers and methods classes I teach" (SR 85). Clearly many mathematics teacher educators are still focusing on this issue.

Research indicates that some educators have challenged the American Association of University Women's (AAUW) report, *How Schools Shortchange Girls* (1992). For example, Kleinfeld (1998) explains why people continue to believe that schools are not appropriately serving the needs of girls despite the apparent advantage they have over boys with regard to grades, discipline, and promotion. Her initial analysis of the issue prompted her to ask professors and undergraduate students whether or not they agreed with the assumption that men perform better than women with regard to schoolwork. She found that a disparity in views existed in "the far right hand tail of the normal curve" in an unrepresentative population of top-tier schools in which males were perceived to perform better than females. Compared to women, men are believed to be more variable; that is, their range of academic and physical capability is larger, which accounts for more males than females in the numbers of mathematically successful males and males in special education. Kleinfeld uses this variability hypothesis to explain the widely held perception that schools are shortchanging girls.

Although the variability hypothesis seems to explain why so many men are outperforming women in mathematical and scientific domains, this hypothesis does not explain why females lead in terms of verbal ability and writing skills throughout years of school. If men are truly more variable, this hypothesis should extend to domains other than mathematics.

Kleinfeld (1998) also notes areas in which males are still ahead of females: "males out-rank women at competitions at the top—SAT examinations, Graduate Record Examinations, the stars and prize-winners in a field of endeavor" (1998, p. 74); such results are consistent with reports of achievement on most high-stakes tests (Educational Testing Service, 2001; National Center for Education Statistics, 2000). In contrast, the National Coalition for Women and Girls in Education (NCWGE) found that females consistently receive better grades on report cards than males (NCWGE, 2002). However, on the National Assessment of Educational Progress (NAEP) the gap between males and females in mean mathematics scores has narrowed, but differences appear when considering student affect between genders. For example, "Gender differences in students' agreement with the statement 'I am good at mathematics' were marked and significant at each grade level. In particular, only 59% of 4th-grade girls reported that they were good at mathematics in contrast with 70% of boys" (Lubienski, McGraw, & Strutchens, 2004, p. 326). This difference in percentage agreement was evident across grade levels and marked a decline for females since 1990 (Lubienski et al., 2004). Thus, although females now score similarly to males on NAEP, they continue to lack confidence in their abilities.

These disparities seem to be an indicator that discussion about gender equity is far from over. In a society that equates achievement in science and mathematics with success and general intelligence, gender equity is still an issue to consider and research.

Theme #2: Lack of Standardization Regarding Equity Issues

The survey responses indicate a lack of agreement about topics to be included in a secondary mathematics methods course. Responses clustered around two issues: who decides what topics should be included in courses, and in which course(s) should gender equity be included.

Teacher educators have considerable autonomy in determining the topics to include in the courses they teach. Many base these choices on their passions as well as on topics that have been important in their educational careers. Four respondents to the survey noted that the inclusion of gender equity depends on an instructor's research interest and this interest was the primary reason they had chosen to include gender equity in their course. One individual, "[being] previously a dean at an all girls school which explored the issue of gender equity in mathematics and science" (SR 18), integrated the experience into the classroom. Many catered their courses to the wants of their prospective undergraduate students or the needs of the community. For example, "Unfortunately my [undergraduate] students wanted more technology so I had to sacrifice something" (SR 39) or "It is a very important issue but things like classroom management and how to survive in the classroom seem more important to them" (SR 74) or "Locally, more important is equity with respect to socio-economic status and ethnicity" (SR 58). Perhaps teacher educators are unable to make unbiased decisions about their curricula because of the outside influences they face. If the curriculum for future teachers depends on the instructor, how can instructors be sure that prospective teachers are adequately prepared to teach in the varied and diverse communities they might encounter?

Approximately 12% of the respondents suggested that gender equity was an important topic but would be covered in general courses that examine multiculturalism and diversity in education rather than in secondary mathematics methods courses. Are there particular gender issues which might be more appropriate to discuss in a mathematics context than in a general education context? Although gender equity might be approached in general education courses, the survey indicated that some faculty members were not aware of these possibilities. Almost 20% did not know if a multiculturalism course was offered at their institution. Of those who were aware of a multiculturalism course offering, some did not know the topics addressed in the course. One respondent stated, "Our [undergraduate] students are required to take a course in multiculturalism but I don't know what content is covered" (SR 60). These results raise two questions. Should equity, specifically gender equity, be discussed in more than one course? Should mathematics teacher educators be aware of whether gender equity is taught in another course?

Would approaching a subject from a variety of standpoints and courses benefit preservice teachers? Do teacher educators expose preservice teachers to multiple perspectives or experiences so that they can draw their own conclusions? Including gender equity in multiple courses could give preservice teachers a wide perspective of the issue and a chance to avoid weaknesses in perspectives from only one course. For example, a review of commonly used textbooks in education courses indicates that many have shortcomings with regard to gender equity (Zittleman & Sadker, 2002). Gender issues receive more attention in foundation and introductory texts than in methods courses, so these classes may provide an appropriate setting to study the topic (Zittleman & Sadker, 2002). Inclusion of gender equity in foundation and introductory courses would allow preservice teachers to consider many sides of the debate, such as how schools are shortchanging boys as well as girls. Also, a multicultural education course places gender equity in a context of other types of bias and underachievement.

A mathematics methods course also seems an apt setting for such a topic so that preservice teachers can closely examine issues related specifically to secondary mathematics. However, elementary and secondary methods textbooks allotted only 0.6%[i] of their pages to gender-related issues, in addition to other gender biases such as illustrations portraying teachers as female and males as administrators, technology users, and special education students (Zittleman & Sadker, 2002).

Perhaps, standardization is not necessary in terms of deciding in which courses to include gender equity. Nevertheless, it is imperative that preservice teachers are encouraged to investigate the issue at multiple points in their educational experience. Gender equity is addressed to a different degree in each textbook, so exposing preservice teachers to many courses that investigate gender equity will provide a more complete perspective than any one course alone. Communication and coherence within an institution would also eliminate the need to guess where a particular topic, such as equity, is discussed, ensuring that preservice teachers understand the effects equity has on teaching.

The importance of preparing teachers to address equity is included in the performance-based standards for professional certification developed by the Council of Chief State School Officers (CCSSO, 1992). Teachers should be able to expect and appreciate individual differences, help students develop self-confidence, and have high expectations for all students (CCSSO, 1992). It may be time for a professional organization, such as AMTE, to provide guidelines for specific standards suitable for prospective mathematics teachers. In NCTM's *Principles and Standards*, a coherent curriculum is defined as one that "effectively organizes and integrates important mathematical ideas so that students can see how the ideas build on, or connect with, other ideas, thus enabling them to develop new understandings and skills" (NCTM, 2000, p. 14). A connected and coherent curriculum is equally important for mathematics teacher education.

Theme #3: Incorporation of Gender Equity into Curricula for Prospective Teachers

Several obstacles arise in determining how best to approach the teaching of gender equity. Many undergraduate students do not see it as an important issue. For example, "My current [undergraduate] students (including females) do not see gender equity to be as big a problem as it once was, so it can be difficult to sensitize them to the issues involved (many of the issues are subtle ones)" (SR 21) or "They all come to my methods class knowing how important the issue is" (SR 66). Preservice teachers with different attitudes towards an issue provide a challenge for posing useful questions and sparking discussion.

The survey found that 65.9% of respondents used a textbook in their secondary mathematics methods courses; of those, 17.9% included the NCTM *Principles and Standards* (NCTM, 2000) as a reference. The textbooks vary in their approach to gender equity: some make no mention of equity; others mention equity but not specifically gender equity; still others provide 1-2 page summaries of research in gender equity with global recommendations and points for discussion. In addressing gender equity, one book offers the following suggestions: "A female engineer speaking to a class can serve as an excellent role model to help girls recognize that even male-dominated careers are a possibility for them. Some teachers also assign career projects to mathematics students" (Brahier, 2004, p. 316). Another includes the following question for discussion: "What are some specific tactics that you as a teacher can use to encourage gender equity?" (Huetinck & Munshin, 2004, p. 357). The *Principles and Standards* provide a vision for equity, but not prescriptive suggestions for what prospective teachers should consider when entering a classroom.

In addition to the texts, some respondents assigned readings related to gender equity, such as those in NCTM's *Changing the Faces of Mathematics* (2000, 2001) series. Among the respondents, 7.1% had supplemented from this series and 24.7% provided citations for other articles they had used (see Appendix B). Hence, it appears that mathematics teacher educators are relying on texts for their sources of gender equity or using alternative readings.

Some respondents expressed concern with relying on such readings for a proper investigation of the issue and suggested alternate ways to address gender equity, including activities such as classroom research, in-class discussion, reflecting on lesson plans with equity issues in mind, and acting out skits. As one respondent asserted, "It takes more than readings to educate teachers. Our undergraduates do 'classroom research' in the field. They collect data, read, and discuss equity issues with secondary teachers and then they write about it" (SR 56). A hands-on approach is one way to make the issue meaningful. Another respondent suggested raising awareness through "in-class discussions rather than through reading articles" (SR 75). Another suggested incorporating gender equity into practice: "We do a lot of emphasizing the integration of gender issues with preparing for instruction, implementing instruction and reflecting on instruction. Our supervisors typically do specific observations to provide data to the student teachers about gender interaction that are observable in stuident (sic) teaching" (SR 82). One respondent stated that her undergraduate students enjoyed the micro-inequity skits from MAA's *Winning Women Into Mathematics* by Patricia Clark Kenschaft (1991). This compilation of skits illustrates the subtle inequities that exist in a secondary classroom. These activities suggest no shortage of creative ways to infuse gender equity into the curriculum. However, perhaps it is time for members of AMTE to provide guidance and collective resources to share with colleagues.

Conclusion

As a community, teacher educators must acknowledge the need for education on the issue of equity, specifically gender equity. Guidelines and principles would make goals clear and consistent and aid teacher educators in making decisions about what to teach and how best to teach it. If standardization with regard to gender equity is successful, it can serve as a model for addressing other biases. An equitable education for males and females can pave the way for equitable education for other underrepresented groups.

The following questions are recommended starting points to guide the mathematics teacher education community in continuing conversations about the inclusion of equity in prospective teacher curricula:

Curricular Issues

- Should gender equity be standardized in courses for prospective secondary mathematics teachers?
- Should gender equity be taught in a methods course? In a foundations course? How might it be approached differently if taught in both? (See Table 1.)
- What would make the curriculum for prospective secondary mathematics teachers more connected and coherent than at present?

Resources/Materials

- What resources best facilitate an investigation of gender equity?
- What activities are the most powerful in sensitizing prospective teachers to the issues of gender equity? Or equity in general?

Issues for the Prospective Mathematics Teacher

- What do grades in school versus standardized test scores tell us about the mathematical preparation of an individual or success in their chosen field?
- Why do some people think that gender equity is important whereas others do not?
- What are society's current views of mathematics education?
- How does gender interact with other forms of bias in the classroom, such as socioeconomic status, race, or ethnicity?
- What can be done to improve all students' attitudes towards mathematics?

Table 1. Advantages and Disadvantages of Gender Equity Taught in Different Courses

	Foundations Course	Mathematics Methods
Advantages	• Equity can be examined across multiple disciplines, such as mathematics, science, and humanities. • Preservice teachers are encouraged to form their own opinions with regard to equity. • Preservice teachers can examine their personal experience.	• Look in detail at mathematics-specific issues and perceptions. • Direct solutions to equity problems are discussed (e.g., increasing wait time, reviewing own classroom practices). • Given that many methods courses include a field placement, preservice teachers can keep equity in mind while observing teachers before their own student teaching begins.
Disadvantages	• Equity can be ignored. • Preservice teachers may miss points that are relevant to the mathematics classroom. • What is included depends on the instructor.	• Equity can be ignored. • How the issues are viewed and taught depends on the instructor. • Preservice teachers may blindly adopt instructor's viewpoints without further analysis. • Preservice teachers may be influenced by more general views of equity in previous classes and not apply equity to the mathematics context.

References

American Association of University Women. (1992). *How schools shortchange girls: A study of major findings on girls and education.* The Wellesley College Center for Research on Women: AAUW Educational Foundation.

Brahier, D. (2004). *Teaching secondary and middle school mathematics.* Boston: Allyn & Bacon, Inc.

Council of Chief State School Officers. (1992). *Model standards for beginning teacher licensing, assessment, and development: A resource for state dialogue.* Retrieved May 10, 2005 from http://www.ccsso.org/content/pdfs/corestrd.pdf.

Educational Testing Service. (2001). *Differences in the gender gap: Comparisons across racial/ethnic groups in education and work.* Princeton, NJ: Author.

Huetinck, L., & Munshin, S. N. (2004). *Teaching mathematics for the 21st century: Methods and activities for grades 6-12.* Upper Saddle River, NJ: Prentice Hall.

Jacobs, J., Becker, J. R., & Gilmer, G. F. (Eds.). (2001). *Changing the faces of mathematics: Perspectives on gender.* Reston, VA: National Council of Teachers of Mathematics.

Kenschaft, P. (Ed.) (1991). *Winning women into mathematics.* Washington DC: Mathematical Association of America.

Kleinfeld, J. (1998) Why smart people believe that schools shortchange girls: what you see when you live in a tail. *Gender Issues, 16*(1-2), 47-64.

Lubienski, S. T., McGraw, R., & Strutchens, M. E. (2004). NAEP findings regarding gender: Mathematics achievement, student affect, and learning practices. In P. Kloosterman (Ed.), *Results and interpretations of the 1990-2000 mathematics assessments* (pp. 305-336). Reston, VA: National Council of Teachers of Mathematics.

National Center for Education Statistics. (2000). *Trends in educational equity for girls and women.* Washington DC: U.S. Government Printing Office.

National Coalition for Women and Girls in Education. (2002). *Title IX at 30: Report card on gender equity.* Washington DC: U.S. Government Printing Office.

National Council of Teachers of Mathematics. (2000). *Principles and standards for school mathematics.* Reston, VA: Author.

______. (1998). *The Mathematics education of underrepresented groups.* Retrieved April 14, 2005 from http://www.nctm.org/about/position_statements/position_statement_05.htm.

Posamentier, A. S. , & Stepelman, J. (2002). *Teaching secondary school mathematics: Techniques and enrichment units (6th edition).* Upper Saddle River, NJ: Prentice Hall.
Secada, W. G. (Ed.). (2000). *Changing the faces of mathematics: Perspectives on multiculturalism and gender equity.* Reston, VA: National Council of Teachers of Mathematics.
Zittleman, K., & Sadker, D. (2002). Gender bias in teacher education texts: New (and old) lessons. *Journal of Teacher Education, 53* (2), 168-180.

---

[i] For comparison purposes, topics such as "Using rubrics to evaluate student work" (Posamentier & Stepelman, 2002, pp. 158-170) comprised approximately 12 pages (or 6% of the pages addressing issues); "Administering a test" (Posameniter & Stepelman, 2002, pp. 183-4) with subheadings "Alternatives for administration", "Classroom arrangement", "Alertness during proctoring", "Early Finishers", and "Absentees" used approximately 1.25 pages (or 0.6%). We chose Posamentier & Stepelman (2002) because it was an edition of a text selected for the research in the Zittleman & Sadker (2002) publication.

Lauriann Kress, a 2005 graduate from Bucknell University with a B.S. in Mathematics and 2006 M.A. graduate from Teachers College, Columbia University, is a member of the 2005 cohort of Newton Fellows at Math for America (www.mathforamerica.org).

M. Lynn Breyfogle is an Assistant Professor of Mathematics and Mathematics Education at Bucknell University. She enjoys investigating issues and topics related to K-12 mathematics teacher education and professional development.

Appendix A
On-line Survey Regarding Gender Equity in Mathematics Teacher Education Curriculum

1. Have you taught a secondary mathematics methods course in the past two years?
   a. I have taught a course.
   b. I haven't taught a course.

2. Did you use a textbook in this course? Yes No

3. Which one(s)?
   a. Did not use textbook
   b. Cangelosi, J. *Teaching Mathematics in Secondary and Middle School: An Interactive Approach*. Prentice Hall.
   c. Cooney, T. *Mathematics, Pedagogy, and Secondary Teacher Education*. Heinemann.
   d. Posamentier, A. et al. *Teaching Secondary Mathematics*. Prentice Hall.
   e. Brahier, D. *Teaching Secondary and Middle School Mathematics*. Allyn & Bacon, Inc.
   f. Brumbaugh, D. *Teaching Secondary Mathematics through Applications*. Lawrence Erlbaum Associates, Inc.
   g. Other: ________________________________

4. Did you supplement the course with outside readings from NCTM's (2001) *Changing the Faces of Mathematics: Perspectives on Gender*? Yes No

5. Which chapters did you use?
   a. Did not supplement from *Perspectives on Gender*.
   b. Fox, L., et al. "Psychosocial Dimensions of Gender Differences in Mathematics"
   c. Ahlquist, R. "Critical Multicultural Mathematics Curriculum: Multiple Connections through the Lenses of Race, Ethnicity, Gender, and Social Class"
   d. Tinsley Mau, S., et al. "Powerless Gender or Genderless Power? The Promise of Constructivism for Females in the Mathematics Classroom"
   e. Wilson, P., et al. "Teachers as Researchers: Understanding Gender Issues in Mathematics Education"
   f. Mitchell, C., et al. "Assessing Achievement in Mathematics: Eliminating the Gender Bias"

g. Campbell, P., et al. "Crucial Points in Mathematics Decision Making: Advice for Young Women"
h. Hancock, S. "The Mathematics and Mathematical Thinking of Four Women Seamstresses"
i. Gilmer, G. "Ethnomathematics: A Promising Approach for Developing Mathematical Knowledge among African American Women"
j. Oliver, D., et al. "One White Male's Reflection on Participating in Experiences Related to Gender Equity: An Interview with Dale Oliver"
k. Becerra, A. "Alice and Rachel: Teacher-Leaders of Color"
l. Valdes, L. "A Latina Tale: The Experience of One Latina Mathematician"
m. Buerk, D., et al. "Empowering Young Women in Mathematics through Mentoring"
n. Manvell, J. "Hypatia of Alexandria"
o. Kort, E. "Role Models and Real-Life Experiences: Influencing Girls' Career Choices in Math and Science"
p. McCoy, L. "Remarkable Women of Mathematics and Science"
q. Streitmatter, J., et al. "Girls-Only Classes in Public Schools: Ambivalence and Support"
r. Forgasz, H., et al. "The Social Context and Women's Learning of Mathematics"
s. Perham, B., et al. "Calculate the Possibilities: A Program in Mathematics and Science for Young Women"
t. Keynes, H., et al. "What Can Be Done to Increase Enrollments of Talented Female Mathematics Students? A Study of the Impact of Middle School Intervention Programs"

6. Did you supplement the course with outside readings from NCTM's (2000) *Changing the Faces of Mathematics: Perspectives on Multiculturalism and Gender Equity*?
   Yes No

7. Which chapters did you use?
   a. Did not supplement from *Perspectives on Multiculturalism and Gender Equity*.
   b. Kitchen, R., et al. "Mathematizing Barbie: Using Measurement as a Means for Girls to Analyze Their Sense of Body Image"

c. Damarin, S. "Equity, Experience, and Abstraction: Old Issues, New Considerations"
d. Uecker, J., et al. "Decked Classes: Structuring the Mathematics Program for Radical Heterogeneity"
e. Remillard, J. "Prerequisites for Learning to Teach Mathematics for All Students"
f. Long, V., et al. "Mathematics Education: One Size Does Not Fit All"
g. Bezuk, N., et al. "Successful Collaborations with Parents to Promote Equity in Mathematics"

8. Did you use other supplemental articles to investigate the issue of gender equity? Yes No

9. If so, what are the citations?

10. Does your institution offer a course on multiculturalism that includes gender equity in education? Yes No Not Sure

11. Is it required? Yes No N/A

12. How important an issue do you think gender equity is in mathematics education?
    very important important
    somewhat important not important

13. Are there any comments that you would like to make about the preparation of your teachers with regard to gender equity?

Appendix B
Additional Readings

Becker, J. R. (1995). Women's ways of knowing in mathematics. In P. Rogers & G. Kaiser (Eds.), *Equity in Mathematics Education: Influences of Feminism and Culture* (pp. 163-174). London: The Falmer Press.

Buerk, D. (1985). The voices of women making meaning in mathematics. *Journal of Education, 167* (3), 59-70.

Chappell, M. F., Choppin, J., Salls, J. (Eds). (2004). *Empowering the beginning teacher of mathematics in high school.* Reston, VA: National Council of Teachers of Mathematics.

Fennema, E. (2000). *Gender and mathematics: What do we know and what do we need to know?* Paper presented at the National Institute for Science Education Forum, Detroit.

Fennema, E., Carpenter, T. P., Jacobs, V. R., Franke, M. L., & Levi, L. W. (1998). A longitudinal study of gender differences in young children's mathematical thinking. *Educational Researcher*, *27* (5), 6-11.

Fennema, E. (1990). Teachers' beliefs and gender differences in mathematics. In E. Fennema & G. Leder (Eds.), *Mathematics and gender* (pp. 169-187). New York: Teachers College Press.

Jacobs, J. E., & Becker, J. R. (1997). Creating a gender-equitable multicultural classroom using feminist pedagogy. In J. Trentacosta & M. Kenney (Eds.), *Multicultural and gender equity in the mathematics classroom: The gift of diversity* (pp. 107-114). Reston, VA: National Council of Teachers of Mathematics.

Kenschaft, P. (Ed.). (1991). *Winning women into mathematics*. Washington DC: Mathematical Association of America.

Roeder, J., & Simms, J. *An activity based on a gender equity dialogue.* Available online: http://www.woodrow.org/teachers/math/gender/05dialogue.html

Sanders, J., Koch, J., & Urso, J. (1997). *Gender equity sources and resources for education students*. Mahwah, NJ: Lawrence Erlbaum Associates, Inc.

Sutton, J., & Krueger, A. (Eds). (2002). *EDThoughts: What we know about mathematics teaching and learning*. Reston, VA: National Council of Teachers of Mathematics.

Tobias, S. (1995). *Overcoming math anxiety*. New York: W. W. Norton & Company, Inc.

Trentacosta, J., & Kenney, M. J. (Eds.). *Multicultural and gender equity in the mathematics classroom: The gift of diversity.* (1997 Yearbook.) Reston, VA: National Council of Teachers of Mathematics.

Zittleman, K., & Sadker, D. (2003). Teacher education textbooks: The unfinished gender revolution. *Educational Leadership*, 60, 59-63.

Luebeck, J. L.
AMTE Monograph 3
The Work of Mathematics Teacher Educators
©2006, pp. 79-95

6

Linking Teachers Online: A Structured Approach to Computer-Mediated Mentoring for Beginning Mathematics Teachers[1]

Jennifer L. Luebeck
Montana State University – Bozeman

It is generally understood that beginning teachers need support and guidance as they begin their careers. Evidence suggests that mentoring and induction can increase teacher effectiveness, promote reflective practices, and positively affect teacher retention. In addition, beginning teachers can work on acquiring, deepening, and applying content and pedagogical content knowledge—but this requires a mentor experienced in the discipline. Content-based mentoring poses a special challenge in isolated rural areas. This paper reports on a program that provides "virtual" content-based mentoring to beginning mathematics teachers in a computer-mediated environment. Successes, challenges, results of early research, and ongoing adaptations are discussed.

Developing the art and science of teaching mathematics is an ongoing and cumulative process. Mathematics educators prepare preservice teachers in the best possible ways, aware that teaching students "for real" far surpasses any preparation provided through exposure to content, methods courses, and a varied menu of field experiences. As teachers mature, mathematics educators promote increased content knowledge and deeper understanding of the "art" through a myriad of creative professional development experiences. Both preservice and inservice teachers benefit from innovative strategies, models, and mechanisms for delivering the content knowledge and pedagogical craftsmanship that teachers need to know.

---

[1] This material is based upon work supported by the National Science Foundation under Grant No. DUE: 0227184 entitled Electronic Mentoring for Student Success (eMSS). Any opinions, findings, and conclusions or recommendations expressed in this material are those of the author and do not necessarily reflect the views of the National Science Foundation.

Unfortunately, little attention has been paid to the process of beginning teacher induction as it relates to mathematics education. Feiman-Nemser (2001) emphasizes that successful teaching practice requires coherent and sustained teacher development from preservice preparation through the early years of teaching. Beginning teachers need support and guidance as they begin their journey toward professional excellence (Feiman-Nemser, 2001; Luft & Patterson, 2002). However, the nature of this support varies widely depending on a school district's size, philosophy, and financial resources. On the basis of these factors, a beginning teacher's transition into a first teaching experience may receive minimal external support.

A well-designed induction program can increase beginning teachers' effectiveness during the early years of their career (Weiss & Weiss, 1999). In particular, well-facilitated induction can promote teacher self-reflection about practice (Britton, 2003; Feiman-Nemser, 2001), which "can lead directly to improved teaching and learning in the beginning teachers' classroom" (Stansbury & Zimmerman, 2000, p. 5). There is also evidence that induction and mentoring can help reverse attrition rates for beginning teachers, as much as 40 to 50% over the first five years (Ingersoll, 2003). In a review of induction programs, Villani (2002) found high retention rates as a result of two mentoring projects. The University of New Mexico reported that 85% of mentored teachers were still in the classroom after five years, while Montana's Systemic Teacher Excellence Preparation (STEP) Project found 96% of mentored teachers still in practice after three years.

Mentoring by an experienced colleague is a key component of many teacher induction programs. Besides providing emotional support and professional encouragement, a mentor can be instrumental in building content knowledge, broadening the repertoire of pedagogical strategies, and developing cultural awareness, all important aspects of a successful induction program (Darling-Hammond, 1998; Feiman-Nemser, 2001; Luft & Patterson, 2002). Of these benefits from mentoring, it is the acquisition, deepening, and application of content and pedagogical content knowledge that requires more than merely a supportive colleague—a mentor needs to be a supportive and experienced colleague *in the discipline*.

Excellent models of induction abound—new teachers may go through a period of formal training, interact with a group of peers, or be paired with a specific mentor. However, in many schools, induction amounts to being assigned a "buddy" within the school building—someone able to help with classroom management and to negotiate the politics and procedures of a district. What is lacking is someone qualified to support development and application of the skills and knowledge learned during preservice preparation, and to deal with content-related challenges that emerge during the first years of teaching.

Challenges and Solutions

Education systems in areas with low population densities are challenged to provide a meaningful level of induction and mentoring for beginning teachers. In a three-room country school, a new teacher has two colleagues to consult, but time is at a premium when one is serving lunch and the other is acting as principal. Providing content-specific mentoring presents an even greater challenge. A study targeting beginning teachers in the Southwest posited that only 20% of beginning mathematics and science teachers had access to any form of induction program; further, none of these addressed discipline-specific issues of teaching mathematics and science (Luft & Cox, 2001).

Montana's geographic size (4th in the nation), low population density (among the ten least populated states), and proliferation of school districts (well over 400 districts in 56 counties) present daunting demographic and geographic challenges for professional development. A 2003 review by the Northwest Regional Education Laboratory found that nearly 80% of Montana schools can be classified as rural. In many of these schools, a sole secondary mathematics teacher confronts a teaching load that includes grades 9-12, 7-12, or even 5-12, with the nearest content-knowledgeable colleague more than 50 miles away. Over the past decade, Montana educators in mathematics and science have experimented with *computer-mediated mentoring* (CMM) via interactive Internet software, bolstered by occasional face-to-face meetings, to link beginning and experienced teachers. These efforts provided much-needed contact for isolated rural teachers, but did not offer a *curriculum* of structured professional development.

More recently, participation in a NSF-funded Mathematics and Science Partnership has enabled the development of a carefully designed, content-rich, Web-based online mentoring program for mathematics and science teachers across the state. CMM offers the potential to provide quality mentoring for mathematics teachers with a focus on deep understanding of content and related pedagogical strategies. Only recently have teachers' increased access to computers, experience with Internet navigation, and familiarity with Web-based communication made such an opportunity available to most, if not all, educators in Montana. Not surprisingly, CMM has received little attention in the research literature to date. In addition, research that specifically addresses induction for science and mathematics teachers is lacking (Luft, Roehrig, & Patterson, 2002), suggesting that CMM is a field ripe for study.

eMSS Partners and Participants

e-Mentoring for Student Success (eMSS), a Mathematics and Science Partnership funded by the National Science Foundation, is an experimental program designed to explore the feasibility of providing content-based mentoring to beginning science and mathematics teachers in a computer-mediated environment with minimal face-to-face contact. The partnership combines the strengths of the Science Math Resource Center at Montana State University in Bozeman, the New Teacher Center at the University of California in Santa Cruz, and the National Science Teachers Association. Drawing from their combined experience in mentoring and induction, distance education, and teacher enhancement and outreach, the partners are building a sustainable and effective mentoring program that could potentially be reproduced in other states and regions. Although eMSS involves several science disciplines and other states, this paper only addresses outcomes related to mathematics teachers in Montana.

Beyond its institutional partnership, eMSS represents a collaborative effort of university scientists and mathematicians, teacher educators, "veteran" middle school and high school teachers, and those just beginning their teaching careers. eMSS was initiated in Fall 2003, with approximately 70 mentor-mentee pairs (both science and mathematics) participating in Montana and California. In the 2004-2005 academic year, the program grew considerably, and approximately 40 Montana mathematics teachers, many continuing from the previous year, were involved in the program.

Mentors and mentees are matched based primarily on content area and grade level rather than on geographic proximity. Mentors, whose classroom experience ranges from 5 to 30+ years, receive training during a summer institute in basic mentoring skills, in promoting best practices for student learning, and in facilitating quality online dialogue. They continue to develop their mentoring skills, share successes and challenges that surface during interactions with mentees, and prepare for new activities via an online "Mentor Forum" that runs during the school year. Beginning teachers, typically in their first two years of teaching mathematics, are recruited through mailings, administrators, and word-of-mouth, with extra effort to recruit teachers of high-need student populations. Mentees may remain in the program for up to three years.

eMSS Program Components

The timeline for implementation of the eMSS program is aligned with a typical academic year calendar. In 2003-2005, the online activity was supplemented by weekend workshops in spring, summer, and fall; these face-to-face meetings are being phased out as the program becomes even more "virtual" in terms of delivery. The online structure of eMSS provides access to a spectrum of activities and resources made available via WebCT course software and accessible only to teachers enrolled in the eMSS program.

The program provides a variety of avenues for discussing and reflecting on issues of content and pedagogy (see Figure 1). A series of *modules* (recently renamed "inquiries") focused on teacher inquiry and reflection address specific topics, such as classroom environment and management, understanding student learning styles, and linking content to standards. Discipline-specific *content areas* provide Web links, ready-to-use materials, and references as well as mathematics-based discussions and one-on-one contact with a university mathematician and teacher leaders. A variety of facilitated and non-facilitated *discussion areas* allow teachers to share ideas, ask questions of their peers, or raise issues for consideration by the group. Other *resources* and *tools* include state and national standards, links to digital libraries, and a module titled "Getting Started" that helps mentors and mentees become familiar with the site and successfully launch their online relationship.

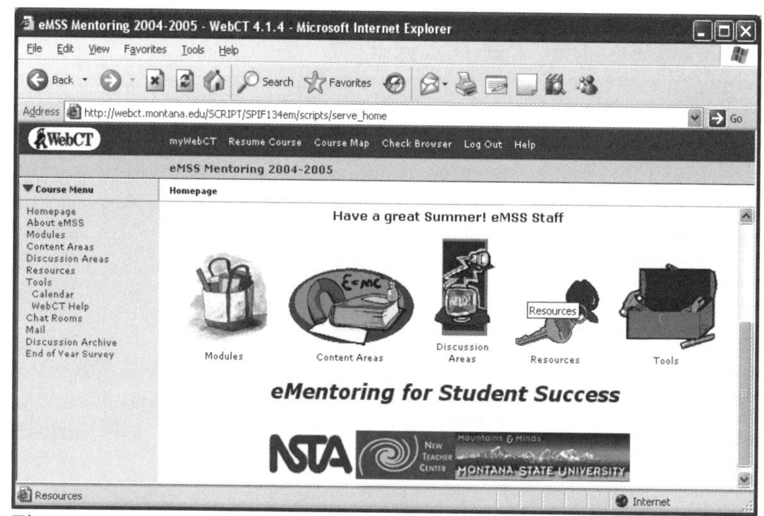

Figure 1. Home page for eMSS mentoring program

Mathematics Content Area

The mathematics content area (see Figure 2) blends archived resources with on-demand information. The ever-expanding archive contains resources for content ("Doing Mathematics") and pedagogy ("Teaching Mathematics") as well as one-click access to standards, Web tools and applets, digital libraries, and professional organizations. The site is maintained, updated, and monitored by a university-level mathematics content consultant (in this case, the author) and one or two facilitators who are exemplary teachers and mentors.

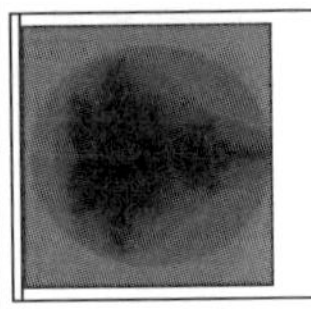

Welcome to the Mathematics Content Area
Greetings to all teachers of mathematics!

In the pages and discussions linked to this site we hope to provide answers to your questions, connect you with resources according to your needs, clarify content, and engage you in conversation about issues in mathematics education, relevant educational reforms, and current mathematics content. We will share ideas for teaching mathematics in a challenging and engaging manner focusing on conceptual understanding and inquiry.

Mathematics experts and teacher leaders are ready to generate discussion and answer your individual questions -- learn about them in "**Meet the Mathematics Facilitators**." Link up to a valuable education database through the "**Math Forum**," or browse several highly rated mathematics resources in "**Mathematics Favorites**." Keep up with the latest advances in "**Mathematics News**," and visit "**Professional Organizations**" to link to national efforts that support mathematics education. Materials related to our content area discussions will be posted in "**Doing and Teaching Mathematics.**" The last item on this menu links you directly to the **eMSS Discussion Area**. The various threads in the "**Mathematics Discussion**" are designed to promote dialogue among mentors, mentees, and facilitators about issues concerning the teaching and learning of mathematics.

The eMSS Mathematics Team hopes you will find these resources valuable for your classroom as well as your own professional growth as a mathematics educator. We look forward to interacting with each of you online!

Figure 2. Navigation page for eMSS mathematics content area

Mathematics Discussion

In an online environment, discourse drives the construction of new knowledge, and eMSS offers several avenues to stimulate discourse about mathematics content, teaching practice, and personal concerns. These include discussion sites for posting announcements, sharing ideas, and debriefing conferences and events; discussions linked to the various teacher inquiry modules; and restricted discussions, such as a site for mentors to deal with issues related to stimulating meaningful dialogue and facilitating eMSS activities.

Dedicated discussions are also assigned to each of the science and mathematics content areas. Beginning teachers are encouraged to pose questions about teaching and learning in the content discussions and to expect responses from mentors, other mentees, the teacher facilitators, or the content consultant. A single query may generate a dozen or more replies as mathematics educators from varying grade levels, traditions, philosophies, and classroom settings comment. Besides initiating and responding to questions, the content consultant facilitates discussion, posts relevant links and information to the content page, and poses problems.

The mathematics discussion interweaves many threads, including discourse related to the content-based inquiry modules, responses to specific questions, and "food for thought" initiated by the content consultant or the teacher facilitators. At times, discussion sites act as a repository for teaching tips and favorite resources, later filtered and posted to the content page archives. Discussion sites also serve as forums for analyzing multiple strategies for presenting a particular mathematical concept. Within any given discussion thread (see Figure 3), ideas are shared, misconceptions exposed and corrected, solutions offered, and encouragement expressed.

Note that Mentee A (see Appendix A) expresses a misconception, which is then further examined through questions posed by Mentors A and B. When Mentee B corrects the misconception, mentors affirm and extend the response. Mentee C's frustrated (and evidently naïve) comments lead the teachers to think about conceptual understanding, consider the use of manipulatives, and share strategies for teaching. Most often questions are raised and resolved within a single discussion thread. However, sometimes a discussion raises enough interest in an issue to warrant further exploration. In such cases, a linked content page can be created to provide additional information, examples, and Web resources.

Teacher Inquiry Modules

One goal of eMSS is to assist beginning teachers of science and mathematics in moving beyond "survival mode" toward a focus on professional practice by encouraging growth in content and pedagogical content knowledge. However, research indicates that successful induction programs match their professional development curricula to the current needs of beginning teachers (Gersten and Dimino, 2001; Rhyms, Allston, and Schulz, 1993). With this is mind, early eMSS activities are focused on basic pedagogical concerns such as classroom management strategies. As beginning teachers gain confidence in the classroom, the emphasis shifts from teacher-centered concerns to addressing pedagogical content knowledge issues that will help beginning teachers effectively deliver content to their students.

Addressing pedagogical content knowledge is accomplished through *teacher inquiry modules*, a series of activities, some classroom-based and some teacher-centered, that combine exploration and reflection. Mentor-mentee partners complete a minimum of two and preferably three modules (one in fall, two in spring), each lasting approximately six weeks. Topics range from early-stage professional issues (e.g., managing student behavior, developing sound classroom procedures) to more advanced inquiries on assessing student understanding, differentiated learning, and diversity. In Fall 2005, first-year teachers had three choices of inquiries: Managing Student Behavior I, Classroom Procedures, and Parent Communication. Beginning teachers beyond their first year chose from four inquiries: Managing Student Behavior II, Classroom Procedures, Design Challenge, and Using Data in the Classroom.

Each year, participants must complete at least one *content-based inquiry* which encourages thoughtful planning, pre-assessment of students, and gathering of resources in preparation for teaching a specific unit or concept. (In the list above, "Design Challenge" and "Using Data in the Classroom" are content-based.) Appendix B displays a portion of the instructions for the first activity in a content module from 2004. As with all teacher inquiry modules, the activities are highly interactive. Participants are expected to locate information, confer with their mentor partners to develop ideas, and share them in group discussion.

Pair Place

A final interactive alternative is Pair Place, a venue for each mentor-mentee pair to exchange messages privately. Although other discussion areas are facilitated with the intent to encourage and guide discourse, Pair Place is purposely unstructured, allowing partners to develop their own best use of the online environment. Preliminary findings (Bice, Simonsen, & Luebeck, 2005) suggest that this private paired discussion area is an effective venue for trust-building, support in matters of life and logistics, and sharing of general pedagogical knowledge. Questions and concerns based on content or pedagogical content knowledge are generally directed to the mathematics discussion area and receive multiple responses.

Preliminary Findings

Research Efforts

Inquiry into the quality, influence, and long-term effectiveness of eMSS is being conducted by a variety of research entities. Assessment of program outcomes is included in the agenda of the in-house research team at the New Teacher Center in Santa Cruz and researchers affiliated with the NSF Center for Learning and Teaching in the West at Montana State University. External evaluators are conducting dialogue analysis of online discourse in order to assess the quality of the discourse in relation to project goals (e.g., deepening content knowledge and broadening pedagogical content knowledge). Program facilitators (including the author) are conducting content analyses of program components to trace the construction of new knowledge about mathematics content and instruction.

Early outcomes

As of 2005, eMSS was in its third year of implementation, with most initial participants still engaged in the program. It is too early to measure the program's effects on teacher performance and retention, but some formative data is available to describe levels of participation in program components, perceived usefulness of the inquiry modules, and influence on content knowledge vs. pedagogical skills. Although the program includes California and will pilot a multi-state approach in Year Four, data reported here include only Montana teachers in Year Three.

As of October 2005, 33 first- or second-year teachers were engaged in eMSS (teachers in their third year are no longer classified as "beginning"). In the first ten weeks, each participating mentee had posted an average of 30 or more messages and read two to three hundred messages (with significant variance between frequent posters and "lurkers" who post rarely). Mentors and mentees post with roughly equal frequency (56% and 44%, respectively). More than 40% of those postings have occurred in the private Pair Place discussions, with 20% occurring in the content areas. The remaining postings are spread among pedagogy-based discussions, mentor discussions about facilitation, and non-specific social conversation. A deeper analysis of Pair Place interactions found that 63% of Montana mathematics teachers meet or exceed program expectations (e.g., they post at least weekly and response "lag time" is minimal; a significant percentage of postings focus on the inquiry modules; and there is evidence of a developing trust relationship).

Mentors and mentees completed pre-post questionnaires in Fall 2003 and Spring 2004. After one full year with eMSS, mentees rated themselves as more prepared in the category of "Basic Teaching/Management Skills." Although their preparedness in the overall category of "Content-Specific Pedagogical Skills" did not increase, the ratings in one subcategory—identifying and developing lessons aligned to address student needs—increased from 46 to 76% (Ford, 2005). Because one content-based eMSS inquiry module addressed lesson design, this change could be the result of mentee participation in that module.

Analysis of formative data (interviews, observations, and analyses of on-line dialogue) suggests that mentors and mentees view eMSS primarily as a source for resources and advice (e.g., activities, lesson plans, useful websites, and opportunities to communicate with experienced teachers and/or content experts). The mixed results regarding effects on basic teaching and management skills vs. content-specific pedagogical skills suggest that greater emphasis needs to be placed on creating opportunities for mentees to think deeply about content.

Challenges and Solutions

Content vs. Pedagogy. The research findings from 2004-2005 suggest that eMSS' next challenge is to "*expand* participants' expectations of their role and the site, to provide training and support for mentors and discussion group leaders in this effort, and....To develop strategies that will ensure an appropriate balance between addressing the immediate needs and concerns of the new teacher and working to help them develop the knowledge and skills of an accomplished teacher" (Ford, 2005, n.p.).

These concerns were addressed for 2004-2005 through increased mentor training, online support in facilitation skills, more content-based teacher inquiries, and development of consistent and user-friendly eMSS content areas. Analyses comparing the preparedness of mentees who participate in particular modules with those who do not will be undertaken in the future. Similarly, plans are in place to assess the impact of the content-focused aspects of the program on the content preparedness of the mentees (Ford, 2005).

Participation. Teachers are not required to post messages in the content areas; when they are engaged in "mandatory" eMSS activities, the mathematics discussion diminishes. However, they are reminded via weekly email to comment on new discussion topics and explore newly posted resources (see Figure 3). As with any professional development effort, there are those who plunge whole-heartedly into the experience and others who remain on the fringes. Overall, at least 60% of the mathematics mentors and mentees post regularly (weekly or more often) to the mathematics discussion.

Subject: Your input needed!

Author: Math Facilitator

Date: Tuesday, October 25, 2005 1:03 pm

Greetings! Three little things for you today...

1. It was GREAT to see you all in Missoula [at the annual teachers' conference] last week--both greeting old friends, and meeting new mentees. What a wonderful group!
2. We have a "new" MATH DILEMMA posted. It's really a revision of the previous one, split into two parts so we can carry that discussion further. Please check!
3. I NEED HELP!!! My secondary methods students are asking about how to award partial credit. I've started a thread on this topic in the math discussion. Please jump in with ideas from your own experience! They'll believe you more than me -- :)

Figure 3. E-mail invitation from content facilitator

Time. A key aspect of ongoing adaptation of the eMSS program components is an effort to streamline activities and tasks. It is an irony of the profession that beginning teachers have the least time available to indulge themselves in pursuits beyond the day-to-day (indeed, minute-to-minute) demands of their classrooms, just when they need help the most. Mentors have reported feeling frustrated or "deflated" when mentees do not meet their expectations or match their enthusiasm. An effective induction program should minimize distractions from the beginning teacher's primary obligation to teach his/her students. Beyond this, an effective *online* induction program, lacking a concrete "workshop" presence, must also find ways to entice participants to engage in its offerings.

A successful solution—classroom dilemmas. In Year Three, the mathematics content discussion initiated one such enticement through a rotating series of "Classroom Dilemmas." Every two to three weeks, a brief content-related dilemma is posted by a facilitator. Only beginning teachers are invited to respond during the first week; mentors may then join in the discussion. Dilemmas are posed as questions ("Is zero even, odd, or neither? How do you know?") or couched in scenarios such as the example below:

> Hey all, I had this problem "$5 + x < 8$, x is prime, what is x?" for a warm-up and we agreed the answer had to be 2 since it must be less than 3. I said 2 is the only prime number, because negatives also have -1 as a factor and a prime number has exactly two factors. A student then said, "What about -1? It has one and negative one as factors." I didn't know! So, is negative one prime, or is my definition incomplete?

Each dilemma is chosen to meet a set of criteria that include:

- Presentation of a clear and "simple" question related to a specific concept;
- Attention to potential student difficulties and misconceptions;
- Opportunity for teachers to express their own understanding of the concept;
- Potential for multiple solution approaches or methods of justification;
- Modeling of mathematical communication and facilitation of thinking.

Streamlined, user-friendly, and satisfying, Classroom Dilemmas motivate teachers to think deeply about teaching mathematics. They also serve as an inviting portal to other online offerings provided in the eMSS program.

Conclusion

The eMSS vision for online mentoring does not attempt to address all aspects of mathematics teacher induction. For example, beginning teachers in the program may still need a "buddy" in their own buildings or districts to help with logistics and everyday issues. eMSS merges one-on-one mentoring with support from a community of colleagues, and it provides a curriculum of induction into key aspects of the teaching profession. eMSS motivates beginning teachers to think about best practices and encourages them to engage in reflective practice from the outset of their careers. Most importantly, eMSS offers access to resources and expertise in mathematics content and the delivery of that content, including structured mentoring with a *content focus*—an approach that researchers have found to be consistently influential in effective professional development (Garet, Porter, Desimone, Birman, & Yoon, 2001; Kennedy, 1998).

The eMSS model presents the kind of sustained and extended professional development experience that research has revealed to be most effective (Garet et al., 2001; Richardson, 2003; Sparks, 2002) and promotes collaboration among teachers, another key component of successful programs (Garet et al., 2001; WestEd, 2000). Even more important, eMSS opens a window of opportunity to beginning teachers in settings where obstacles of isolation, expense, and competing interests have previously disallowed high-quality mentoring programs. The eMSS model has the potential to alter significantly the early experiences of beginning mathematics teachers who, rather than learning only from their mistakes, now have the means to learn by example and shared experience.

References

Bice, L., Simonsen, L., & Luebeck, J. (2005). *When teachers talk: Online mentoring of beginning mathematics and science teachers.* Manuscript in preparation, Montana State University – Bozeman.

Britton, E., Paine, L., Pimm, D., & Raizen, S. (Eds.) (2003). *Comprehensive teacher induction.* Dordrecht: Kluwer Academic Publishers.

Darling-Hammond, L. (1998). Teacher learning that supports student learning. *Educational Leadership, 55*(5), 6-11.

Feiman-Nemser, S. (2001). From preparation to practice: Designing a continuum to strengthen and sustain teaching. *Teachers College Record, 103*(6), 1013-1055.

Ford, B. A. (2005, April). *E-mentoring for student success: Year three evaluation report.* Chapel Hill, NC: Horizon Research, Inc.

Garet, M. S., Porter, A. C., Desimone, L., Birman, B. F., & Yoon, K. S. (2001). What makes professional development effective? Results from a national sample of teachers. *American Educational Research Journal, 38*, 915-945.

Gersten, R., & Dimino, J. (2001). The realities of translating research into classroom practice. *Learning Disabilities Research and Practice, 16*(2), 120-130.

Ingersoll, R. M. (2003). The teacher shortage: Myth or reality? *Educational Horizons, 81*(3), 146-52.

Kennedy, M. M. (1998). *Form and substance in in-service teacher education.* (Research monograph No. 13). Arlington, VA: National Science Foundation.

Luft, J., & Cox, W. (2001). Investing in our future: A survey of support offered to beginning secondary science and mathematics teachers. *Science Educator, 10*(1), 1-9.

Luft, J., & Patterson, N. (2002). Bridging the gap: Supporting beginning science teachers. *Journal of Science Teacher Education, 13*(4), 267-282.

Luft, J., Roehrig, N., & Patterson, N. (2002). Barriers and pathways: A reflection on the implementation of an induction program for secondary science teachers. *School Science and Mathematics, 102*(5), 222-228.

Rhyms, R., Allston, D., & Schulz, L. (1993). Effective teaching does make a difference. *The Alberta Journal of Educational Research, 39*(2), 191-203.

Richardson, V. (2003). The dilemmas of professional development [Electronic version]. *Phi Delta Kappan, 84,* 401-406.

Sparks, D. (2002). *Designing power professional development for teachers and principals.* Oxford, OH: National Staff Development Council.

Stansbury, K., & Zimmerman, J. (2000). Lifelines to the classroom: Designing support for beginning teachers. A WestEd Knowledge Brief. San Francisco: WestEd.

Villani, S. (2002). *Mentoring programs for new teachers*. Thousand Oaks, CA: Corwin Press.

Weiss, E. M., & Weiss, S. G. (1999). Beginning teacher induction. (ERIC Document Reproduction Service No. ED 436 487)

WestEd. (2000). *Teachers who learn, kids who achieve: A look at schools with model professional development* [Electronic version]. San Francisco: Author.

Jennifer L. Luebeck is an assistant professor of mathematics education at Montana State University in Bozeman. Her work on campus is primarily with secondary preservice teachers; she also designs and teaches distance courses for inservice mathematics teachers in master's- and doctoral-level mathematics education programs at MSU. Her research and professional interests blend distance learning with designing and delivering effective models of professional development in mathematics, particularly for beginning teachers and teachers in isolated rural areas.

Appendix A
Excerpt from an eMSS mathematics discussion thread

Facilitator: Explain in your own words how fractions and decimals are related.

Mentee A: I'll be brave and start this one out. A decimal is a method of expressing any fraction as if it were in hundredths. It is more precise than a fraction and makes it easier to add and subtract since everything is in the same unit, same denominator, equivalents. I admit this will break down and require more explanation when you get to decimals less than .01 but it's a start eh?

Mentor A: Is a decimal really more precise than a fraction?

Mentor B: I was wondering about that because I was thinking of repeating decimals -- are they more accurate when they are written as a fraction or are they the same because you can just use the bar over the top of the number? If you are trying to figure the cost of a TV that is 1/3 off, it really changes my discount by multiplying out more places (multiplying by .3 compared to multiplying by .3333). Does that mean one is more accurate than another?

Mentee B: To add to the discussion on "is a decimal really more precise than a fraction?" I always tell my students to try and warm up to the idea of fractions because they are more exact. A decimal can be expressed in a variety of ways (tenths, hundredths, etc.) but rounding is generally involved. This is an approximate answer and less than exact....It is easier to perform basic calculations (add, subtract, multiply, divide) on decimals since they are in the same form.

Mentor C: You are right, Mentee B. Many decimals are approximations and any time you approximate, you lose something in the answer. Using the fraction is exact, but it is sometimes not as easy for students to manipulate as a decimal. Mostly it depends upon what you want for a solution. If you want precise, you use fractions; if you want approximate, you use decimals. The students really need to know how to do both and know the reasons why.

Mentee C: Fractions are just undivided decimals. Example, the fraction 1/4 I read as "one divided by four" which if you did out "long hand" is .25.

Mentor D: Another connection between fractions and decimals....If the value is first given as a decimal, and the decimal is read correctly, then the fraction form is also stated. For example .35 is read "thirty-five hundredths," and in fraction form is thus written 35/100.

Mentee C: I am perplexed why students have such a hard time with fractions. Operations with fractions take nothing more than knowing a few rules and understanding of addition, subtraction, multiplication, and division. The only thing I see is that operations with fractions take a little more time so the students interpret that as "harder"....My goal is to get kids "fluent with fractions" but it is proving to be a very difficult process. Any ideas on ways to make fractions look easy?

Mentor B: I don't know how to make them easier, but I do have some ideas about what makes them more difficult. All of a sudden some of what they have learned with whole numbers is no longer true—for example: Multiplication makes larger, division makes smaller....When you count there is a next number like after 2 comes 3....The more digits in a number the larger it is....Just some thoughts on why they think it is hard – the "rules" have changed on them.

Mentor A: I wonder if students move way too quickly from the concrete to the algorithm—or maybe they really don't even deal much with the concrete. They need to figure out how to add unlike fractions, etc. and not be given an algorithm so quickly—Maybe???? Fractions do mess with their number sense and that makes it hard.

Mentor D: ….I have found a great "book" resource for teaching fractions, decimals, and percents comes from the following website….Type into the search area "number sense", and it will take you to…the "Discovering Basic Math Concepts, Number Sense" series….Ask for a free 30-day preview.

Mentor E: I'm leaping in here because I'm a fanatic about concepts. I've taken 8th graders (and older) and drawn pictures or folded paper to show what 1/4, 2/3, etc. looks like. Students need to internalize that the denominator tells the "size" of the item, and the numerator "how many" there are. When that is clear, they're much more apt to understand the reason for common denominators and the algorithms that go with them.

Mentee A: YES!! I agree—we've been folding and now we're going into moving and cutting, rods, pictures to make more pictures, etc. More concrete, less abstract, greater understanding…. Part of what makes it difficult is the question of what is the whole? If you're talking about 1/2 of a pizza, that looks a whole lot different than 1/2 of a Cuisenaire rod. I think they need lots more time with the basics, manipulatives and pictures to really get the concepts. That's where I'm heading this week.

Facilitator: I think you make a great point here, Mentee A…it's confusing for students that 1/2 can be different every time and that it depends on the whole. Has anyone ever used 10 X 10 boxes to show the relationships between fractions, decimals and percents? Then you can ask students to think about how they could show the relationships between decimals with values in the 1000th place. How would this change the percent?

Appendix B
Excerpt from an eMSS content-based inquiry module

ACTIVITY 1: Describing the Content You Want Your Students to Learn

Step 1: Introduce yourself. (done)
Step 2: What do you want to know about this topic? (done)

Step 3: Post your "Content Goal."

A. Think about the specific ideas, concepts, and skills that are part of your chosen topic (the one from Steps 1 and 2 that you anticipate teaching a month from now). Share your list with your mentor and come up with one set of ideas within that topic (think concepts, not just facts or procedures) that you think are most important for your students to learn.

B. Reply directly to this message. Describe the set of ideas that you determined are most important for your students to learn. We'll call this your "Content Goal." Explain why you think these ideas are the most important.

EXAMPLE: Your topic for late February is "Right Triangles." You determine that of all the facts, formulas, concepts, and applications related to right triangles, you are most concerned that your students recognize that the Pythagorean Theorem is a relationship that can be proven, not just an empty formula that seems to work all the time. This is your "Content Goal." You feel this is important because you want your students to understand that (1) having proof is different than merely "believing" after seeing lots of examples, (2) proofs don't always have to be "scary" (these will be by pictures and by cutting paper), and (3) they are capable of doing proofs!

Step 4: Compare your "Content Goal" to Standards

A. Look at the Standards to discover what they have to say about your Content Goal.

- National Council of Teachers of Mathematics (select your grade band from the first page, then use the margin to find your content goal): http://standards.nctm.org/document/index.htm
- Montana Math Standards from the OPI Web site (a PDF document you can download and read on your computer): http://www.opi.state.mt.us/PDF/standards/ContStds-Math.pdf

B. Reply directly to this message. Is your "Content Goal" directly or indirectly related to the state or national math standards? Which Standard (NCTM or MT) most closely matches your "Content Goal"? If you can, quote a supporting phrase from the standard(s). When something you plan to teach, or something in a text, is not supported by standards, what should you do?

Crespo, S. and Featherstone, H.
AMTE Monograph 3
The Work of Mathematics Teacher Educators
©2006, pp. 97-115

7

Teacher Learning in Mathematics Teacher Groups: One Math Problem at a Time

Sandra Crespo
Helen Featherstone
Michigan State University

In this paper, we describe a teacher study group model developed from our experience facilitating and researching such groups. This model uses a rich mathematics problem as an anchor to the group's mathematical and pedagogical conversations. Teachers work on the problem, prepare grade-level adaptations to try with their students, and bring artifacts and stories of their students' work on the adapted problems to analyze with the group. We discuss how this sequence of study group activities provides multiple opportunities for teacher learning. We also discuss challenges to consider when initiating and facilitating such groups.

In recent years there has been increased interest in promoting teacher-learning communities as a form of teachers' professional development. This approach engages teachers in critical and reflective conversations about the challenges of teaching practice (Ball & Cohen, 1999) and is founded on collaborative and active views of teacher learning (Wilson & Berne, 1999). These professional learning communities offer "the possibility of individual transformation as well as the transformation of the social settings in which individuals work" (Grossman, Wineburg, & Woolworth, 2001, p. 948). In fact, some argue that school-based professional communities are key to supporting teachers' attempts to change their teaching practices and to implement systemic school reform (e.g., McLaughlin & Talbert, 2001).

Teacher study groups are also consistent with theories of learning as situated, social, and distributed cognition (Putnam & Borko, 2000). These theories lend support to professional development that situates teacher learning in authentic practices of teaching and that engage groups of teachers in intellectual conversations about those practices. A situative perspective on teacher learning argues that "when diverse groups of teachers with different types of knowledge and expertise come together in discourse communities, community members draw upon and incorporate each other's expertise to create rich conversations and new insights into teaching and learning" (Putnam & Borko, 2000, p. 8).

In mathematics education, a growing number of educators have begun to document the complexity associated with forming and sustaining teacher groups (e.g., Arbaugh, 2003; White, Sztajn, Hackenberg & Snider, 2004), reporting on teachers' perspectives and design features that made it possible to sustain their participation in teacher study groups. Others explore the connections between participation in such groups and changes in the participants' knowledge, beliefs and teaching practices (e.g., Kazemi & Franke, 2004) and investigate the possibilities of Japanese lesson study as an approach to supporting mathematics education reform (e.g., Fernández & Yoshida, 2004). In this paper, we describe our approach to teacher study groups and our insights into the opportunities for collective and individual learning that this form of professional development provides to the participants.

The Study Group Model

Our current approach was developed during our most recent work with teacher groups (Crespo & Featherstone, 2003, 2002, 2001). Our model creates opportunities for the teacher participants to explore mathematics together, to learn from each other's stories of practice, and to learn about students' mathematical thinking across the elementary grades. Before illustrating these learning opportunities, we provide an overview of our model by noting specific features of our approach to teacher groups.

1. *School-based.* All participants in each study group teach in the same elementary school. Learning in communities of practice with colleagues from the same school increases the likelihood that teachers will support one another's efforts to teach differently and that they will continue to work together on their practice when the project ends. This feature also faciliates teachers observing in each other's classrooms and collaborating in various ways.
2. *Cross-grade.* Participants teach different elementary grades. This feature creates opportunities for teachers to think about how students' mathematical thinking develops across school grades. It also makes it possible for participants to learn how different grade level teachers and their learners engage with the same task.
3. *Problem-based.* The group discussions about mathematics, students' work, and pedagogy center around a problem carefully designed (or selected) to challenge and extend the teachers'

mathematical understanding while offering possibilities for adaptation for use with elementary aged children.

4. *Focused on participant-generated artifacts*. Participants bring records of practice from their own classrooms that are the object of the group's collective inquiry. This makes it possible for teachers to learn about their own students, and learn to see their own and colleagues' practices as sites for inquiry.

The work of these teacher groups centers on a rich mathematics problem—one that meets many of the criteria for worthwhile tasks outlined in the *Professional Teaching Standards* (National Council of Teachers of Mathematics, 1991), that can engage adults in mathematical inquiry, and that can be adapted to the elementary grade of each teacher in the group. The teachers work on this mathematics problem together, pose it to their own students in their classrooms (making adaptations as needed), and then report at the study group's next meeting on what happened and what they learned from that experience. Our acronym for this sequencing of study group phases of activity is SATRR—*Solving, Adapting, Teaching, Reporting* and *Reflecting.*

More specifically, during the *solving* phase of the SATRR cycle, teachers work together on the problem and discuss their solutions and strategies. The teachers then explore ways to *adapt* the problem for use with their K-5 students and consider what mathematics (in terms of content and processes) their students might learn from a lesson based on a grade-adjusted version of the problem. In the weeks between study group meetings, each teacher *teaches* a mathematics lesson based on the work done in the study group, posing her version of the problem in her classroom, making notes on the class discussion, and collecting samples of her students' work. At the next meeting, the study group members move into the *reporting* and *reflecting* phases. These discussions focus on the teachers' accounts of what happened in their classrooms and an initial analysis of their students' work. The group then reflects on what was challenging, what they learned, and what they would like to try next.

Facilitating SATRR Groups

Engaging teachers in professional discussions of teaching is not without serious challenges. Chazan and Ball (1999), for example, have suggested that most discussions about teaching tend to be too judgmental to support analytical and constructive conversations about practice. Similarly, Pfeiffer and Featherstone (1995) have noted, "when teachers talk about their work, most are quite facile in talking about teaching without revealing the struggles and uncertainties inherent to the practice" (p. 5). Lord (1994) also contends that, in order for teacher groups to become a setting for teacher learning, teachers must be prepared to disagree openly and publicly with practice and to engage in a "critical colleagueship," defined as "an alternative professional stance where teachers move beyond sharing and supporting one another through the change process to confronting practice—the teachers' own and that of his or her colleagues" (p. 193).

The role of the facilitator of study group meetings is, therefore, challenging and complex. The selection of mathematical problems determines the range of mathematical ideas and approaches that become available to the group. The study group leader needs to have a diverse repertoire of problems and be familiar with their mathematical implications and possibilities, for both study group participants and the children they teach. Managing the group's discussion is an equally demanding dimension of the facilitator's work. The voluntary nature of the group, the complexity of norms for politeness and individualism in schools (Little, 1990; Lortie 1975), and the anxiety many elementary teachers feel in relation to mathematics, contribute to the difficulty of the facilitator's role. As Carroll and Thompson (2002) note:

> Finding the right words for "re-voicing" comments demands a blend of genuine curiosity about others' ideas and a tactful command of language to present thoughts in respectful but clear terms. This goes beyond popular conceptions of "active listening" where one might simply re-state what someone else has said. To be effective and helpful for others' negotiation of meaning, the group leaders' re-voicing of comments need to pick up on larger patterns of ideas lurking in the details of the ongoing conversation and rebroadcast them in ways that enable new perspectives or apparent underlying principles to be apprehended or that help people pinpoint their areas of disagreement (p. 24).

In this section, we illustrate a full cycle of the SATRR model using the following "bikes and trikes" problem. With this illustration, we provide some ideas for facilitating the group's conversations. In the subsequent section, we discuss the opportunities this model offers for teacher learning.

> A bicycle and tricycle manufacturer makes bicycles and tricycles with the same size of wheel. He received a shipment of 100 wheels, and he wants to use them all. What combinations of tricycles and bicycles can he

make? Which combinations won't work? Are there any patterns?[i]

Solving

In the solving phase, teachers work on the mathematics problem individually and then share their individual ideas for the group to examine. Invariably, the group's collective understanding of the problem is richer than that of any individual participant. In this particular problem, teachers share different strategies that range from less to more systematic explorations of the combinations that will and will not work. In our experience, some teachers prefer to draw, others make a table, and occasionally someone attempts to construct an equation. As each teacher presents a strategy, the group learns about what each has been noticing. In this particular problem, for instance, teachers notice that only even numbers of trikes will work, that two tricycles can be traded for three bicycles, that one can think of the problem as making groups of five wheels, and that there is an inverse relationship between the number of bikes and the number of trikes. Eventually the group constructs a table and finds all the combinations.

In this phase, the facilitator invites teachers to share and explain their ideas and then to add or respond to what has been said by others. The facilitator presses for elaborations, more details, and greater clarity. In this problem, making explicit what numbers one is referencing is important – is one talking about the number of bikes or bike wheels? The facilitator also asks questions to move the group towards greater understanding: "Are there any solutions we are still puzzling over?" or "How do we know that we have found all the combinations?" When disagreements arise about interpretations related to the problem (e.g., whether or not 0 trikes and 50 bikes is an acceptable combination), the facilitator may ask the group how to settle the disagreement or to consider why the issue related to the disagreement matters and how it might be addressed in their own classroom.

Adapting

After teachers have explored the problem mathematically, they explore the pedagogical potential of the problem for their respective grade levels. Teachers consider how the problem might be adapted for their students and what they might learn. For instance, the primary grade teachers immediately consider reducing the number of wheels in the problem. The older grade teachers, in turn, focus on how to help their students structure their work so that they see patterns and consider whether they would ask students to organize their work in a table.

During this phase, the facilitator invites teachers to analyze the mathematics the students will explore as they work on their new version of the problem. The facilitator probes: What has changed and why? For instance, reducing the number of wheels results in fewer combinations. Therefore the facilitator may press: "how many combinations are too few and how many are too many?" Teachers anticipate the mathematical challenges that their version of the problem will present for their students. As teachers consider how to pose this problem to their students, they discuss the rewording of the problem and how familiar or unfamiliar particular words might be to their students. For example, some teachers wonder whether their students will know the word "trike." Others consider how they can focus their students' attention to the idea that there must be "no left over wheels." Still others wonder whether manipulatives will help or hinder students' mathematical explorations.

Reporting

In the weeks between meetings, the teachers teach their version of the study group problem in their respective classrooms. They come to the next meeting prepared to share what happened. Some teachers bring interesting and surprising stories of what students did; others bring students' work to share with colleagues. The kindergarten teacher in one of our groups brought a video of her lesson to share. Her colleagues, having expressed doubt that kindergarteners could get far with this problem, were amazed to see five-year-olds notice the relationship between the bikes and trikes (e.g., one trike is one bike plus one left over, and one trike and one left over will make two bikes), make predictions, and find all combinations of bikes and trikes given 20 wheels.

Facilitating the reporting phase demands much of the facilitator as this stage can easily turn into a teacher 'show and tell.' Although it is important that participants share successes, it is vital that these are not the only kinds of stories that are told, and that failures and confusions are equally valued. One way to make space for stories of failure is to structure the reporting in particular ways—"share a teaching move that did not quite work out the way you envisioned." Another facilitator move is to ask teachers to share students' work with no interpretation to allow the group to offer insights and questions before hearing the reporting teacher's interpretations (Blythe, Allen, & Powell, 1999). Another is to invite teachers to see the group as a resource to make sense of and figure out whatever is still puzzling them. One teacher, for instance, began her reporting turn by saying, "I wonder if I made the problem too easy for my students" and engaged the group in determining why her version of the problem had become a counting exercise rather than an intellectual challenge for her students.

Reflecting

After each teacher has had the opportunity to report on his or her experience, the teachers move to the reflecting phase, considering what they learned and might want to try next. During this time teachers tend to talk about being surprised by what their students can and cannot do, about the challenge of guiding without leading their students to notice what they want them to notice, and about wanting to do more of these kinds of problems with their students. The teachers reflected for example on the differences and similarities across the grades in how students approached the problem and in common difficulties. At this stage, the facilitator might invite reflection and analysis with questions such as, "What was hard about trying this problem?" and "What did we learn from today's discussions?"

Opportunities for Teacher Learning

These comments from participants suggest the range of learning opportunities that teachers felt the SATRR groups provided.

> I find myself pushing my students to think beyond more. Instead of stopping at solving a problem one way, we talk about alternatives. We also do more thinking/talking about connections to other problems. *(Second Grade Teacher)*

> This group has impacted my thinking and teaching of mathematics in that I have thought more about what kinds of questions I ask the students. Instead of yes/no questions, I am beginning to ask questions that get students to think and discuss more. *(Fourth Grade Teacher)*

In this section, we analyze the learning opportunities that this kind of teacher study group offers to participants. We draw on our own observations and on what teachers have told us about what they have learned by participating in these groups.

Explore mathematics together

Typical interactions among teachers in school settings do not necessarily include spending time exploring mathematics. We were initially concerned that working on mathematics together would make our teachers uncomfortable and perhaps push them into a passive or intimidated stance. Instead, all of the teachers in our study groups appreciated the opportunity to figure out math problems together. The conversations around mathematics problems generated lively and collaborative discussions of solutions and strategies. Opportunities for mathematical discussions also occurred when teachers were adapting problems to make them accessible to their students. In our study groups, we noticed that the teachers' conversations about their own mathematical explorations were in fact more collaborative and exploratory than the talk that followed the teachers' use of the problem in their own classrooms (Crespo, in press; 2002).

Although most teacher study groups are committed in theory to inquiry, there are powerful forces working against the raising of questions that are the *sine qua non* of actual inquiry. Without questions, inquiry cannot occur. However, from where do questions arise? The teacher who tells a story based in her practice makes certain assumptions about the situation because of the fundamental realities of daily life in her classroom. Her capacity to raise questions about her practice is circumscribed by the taken-for-granted nature of her practice. Her listeners have some advantages as question raisers, because they do not necessarily share all of her assumptions about teaching; however, they are often inhibited by the norms of privacy and politeness that shape conversation in groups of teachers (Little, 1990). Furthermore, the teachers often do not know enough about the curriculum of the particular grade of the teacher who is telling the story to formulate good questions.

In SATRR groups, teachers have had the experience of working together on the mathematics in the problem and have helped each other adapt the problem to different grade levels. This prior experience together before sharing about their teaching practice alleviates some of the previous constraints. All of the teachers have worked together on some version of the problem that the teacher is reporting and have analyzed its mathematical features. Most participants have taken the problem into their own classroom and presented their students with some version of it. Having listened to several teachers' accounts of their own teaching of the problem, the group participants have an expanded and refined sense of the mathematical and pedagogical possibilities embedded in the problem. These teachers are likely to wonder about some of what they hear. It is not sufficient, however, for them to wonder if they are silenced by norms of privacy and politeness, by the worry that a question will be interpreted as a criticism. The fact that the lesson reported is a part of the group's joint work has the potential to change these norms. In addition, the teachers, having questioned and challenged one another about how they worked on the math problem during the *Solving* phase of the cycle, can become comfortable asking questions about teaching.

Share stories of practice

In SATRR groups, participants report on what happened in their classes when they tried their version of the problem. These reports provide the participants with multiple learning opportunities. One such opportunity occurs when teachers bring artifacts and stories from their classrooms to share with the group. In describing what they and their students did with the common problem, the teachers invite their colleagues into their classrooms and open their practice to discussion. Another benefit is that teachers take into their classrooms a number of teaching strategies suggested or described by others in the group or generated through group observation or discussions.

In our case, teachers not only shared their practice by bringing stories and students' work, but some of them also brought a video of their teaching; a few others invited the group to their classrooms to observe them try the study group problem. In the first excerpt below, it is clear that watching the kindergarten teacher's videotape leads Brian (a second year teacher) to reflect on the adequacy of the way he introduced the problem to his fourth grade students. It also prompts Marie to think about the role that manipulatives might play in enabling students who could not engage with the words alone to work on the problem productively.

Short excerpt from discussion of Kindergarten teacher video of the bikes and trikes problem:

Facilitator: Comments for Jenny? [Kindergarten teacher] Anything you noticed or questions?

Brian: This was really good. I liked how you introduced the problem. I thought I should have spent more time on that.

Marie: I was surprised that they could figure this out, I had one student in my class that could not figure this out, even when I lowered it to 20. I wonder if having the noodle wheels would have helped him.

Jenny: I'm not sure that my kids could have done this well at the beginning of the year.

Facilitator: So you think that you couldn't have done this earlier in the year?

Jenny: No way, not a chance.

Brian: So what did you do from September to now to get them to this point?

In sharing her presentation of the problem in the following excerpt, Nell offers her colleagues some pedagogical tips: first, they can present the problem as one that their friend faces (which might make it more interesting to the children); second, even vocabulary that seems ordinary to the teacher may confuse this generation of children. In addition, Nell shows several ways of making the problem accessible to the children without diminishing its mathematical richness: counting out the manipulatives ahead of time; offering students a manipulative that *looks* like a wheel.

Short excerpt from discussion of second grade teacher reporting on her experience:

Nell:	I told them my friend called me on the phone and that she's working on a bike shop and that she had this problem: she has 50 wheels and she needs to figure out all the possible combinations of bikes and trikes she could make. Trikes is confusing for some special ed kids and ESL kids so I kept saying tricycles, so we went over the number of wheels that bikes have and that tricycles have. I used the noodle wheels and put them in a bag for them and their partners. I did tell them they should check they should have 50 of them, can't use more than 50 wheels, can't use less than 50 wheels. I really stressed that ALL wheels had to be used.
Facilitator:	So you put 50 in each of them?
Nell:	Yeah, and I did tell them they should check to make sure they did have 50.
Marie:	Did they have any trouble working with 50 wheels?
Nell:	I think every combination was found except for one. I have some samples. Here's Terran, he and his partner are the two that found the most combinations. The way they did it was interesting because they divided their paper in half and labeled one half T and the other N meaning that Terran was in charge of the bikes and Nigel was in charge of the trikes. They were swapping wheels back and forth across the table to each other so they could use them all up, so I thought that was a clever way to do it. You could hear them talking back and forth saying "you need one more," or "I need one

more," they were having some neat conversations together.

Ann: Did they put together 25 bikes and 0 trikes?

Nell: Yeah they did, a few of them did. Here's Tyler, he put the 3 on the top of his chart and 2 for the bikes [...] He's the one who talked a little bit about multiplication when I was writing the combinations on the board. So I am wondering if he kind of noticed that 2 times 25 is 50 and that's why we had a 0 on the trikes? He tried to talk about that but I called on Gina because she had done hers in groups of 10 and in 2nd grade we focus a lot on place value, so when her and her partner were making the combinations they always counted in groups of 10.

By sharing stories and artifacts from their classrooms, teachers invite others to see their classrooms as sites for inquiry. Interestingly, our study group teachers also became comfortable sharing their practice with others who were not part of the teacher groups. At one of the schools, for example, the study group teachers regularly shared during staff meetings what they were doing in their teacher groups. Other teachers who were not in the study group began to incorporate some of the problems from the teacher group into their classrooms.

The teachers in our groups who were also mentor teachers to the elementary teacher education (TE) preservice teachers in our mathematics methods courses became comfortable sharing their practice with the preservice teachers. The study group teachers began to try the study group problems on the days that the TE students came to observe, making it possible for preservice teachers to see children working on problems that they had worked on in their methods classes. These focused observations in their field experiences provide opportunities for the TE preservice teachers to learn important lessons about the teaching and learning of mathematics and to have reflective conversations with their mentor teachers about the challenges of teaching and learning mathematics.

> Mrs. K. took about 5 minutes reviewing the difference between trikes and bikes and quizzing individual children to make sure they were listening. I was surprised at how many children had trouble with the quantity of wheels on bikes and trikes. [...] I guess I assumed that all first-graders had ridden at least one of them in their lifetime, but that's not necessarily true, and even if it were the case, that wouldn't

> guarantee that they could both remember and verbalize the difference. (*Lindsay*)
>
> One of the pairs at my table created a production of only bikes with no trikes. They were the only group to do that. [...] I find this work to be an example of an outstanding response. [...] Mrs. K. never told her students that they could have a situation of all of one kind and none of the other. But this pair was able to discover that if it were solved this way there would be no leftovers. (*Mandy*)

In these statements, TE preservice teachers are learning something new about problem posing and developing admiration for children's mathematical thinking. These insights speak to critical issues in the preparation of teachers. Because most new teachers are white women from suburban and rural backgrounds, cultural differences between TE preservice teachers and children in urban schools often undermine early efforts of TE preservice teachers to engage ALL children with rich mathematical tasks. In the above excerpts, Lindsay realizes (perhaps for the first time) that she cannot assume that her students will have had the out-of-school experiences that she thought were universal and that she must ensure that her students have the background information they will need in order to engage with the mathematics in the task. Another issue in preparing teachers to teach ALL children is that the wider culture 'teaches' prospective teachers to underestimate the intelligence of children, particularly of children they will teach in urban schools. In the previous highlight, Mandy delights in the intellectual achievement and initiative of two of her students—and judges their work as outstanding.

Insights on students' thinking across grades

> I have really enjoyed meeting with the different grade levels and taking a look at the curriculum. It has been very helpful to get a big picture of math K-4. Many underestimate what kindergarteners can do. It has been fun to challenge my students and share with the group how much they can actually do. (*Kindergarten teacher*)
>
> I have gained lots of knowledge about mathematics teaching and problem solving. It is nice to know what the grades before and after mine are teaching. Seeing different perspectives at teaching the same problem has been very beneficial. I hope that we can continue to meet next year. (*Fourth grade teacher*)

Overwhelmingly, teachers identified gaining insights on students' thinking across the grades as a major benefit of participating in cross-grades study groups. By sharing and examining student work that was generated in their own classrooms, the participants analyzed students' ideas not only at their grade level but at other grade levels as well. Many upper elementary teachers were surprised that students in the early elementary grades could engage with sophisticated mathematical ideas when their work was supported and problems were carefully adapted to their level of ability. The primary grade teachers in turn became more attuned to the mathematical implications of their work on number development and counting.

The cross-grade design of this model invites teachers to attend to issues of coherence and continuity of curriculum across the elementary grades. The development of the basic facts from Kindergarten to Grade 4 was one cross-grade investigation that the teachers in one of the groups spearheaded (see Crespo, Kyriakides, & McGee, 2005). This investigation began with a report from a fourth grade teacher that her students were counting on their fingers for the simplest of computations. The group spent several months investigating students' difficulties, collaboratively designing tests, interviews, and teaching experiments with the help of the group's facilitator. This investigation resulted in a report to the school principal and staff based on their insights into students' struggles and curriculum gaps, along with recommendations for implementing approaches other than drill and practice, to teach basic facts across the grades.

Concluding Thoughts

Teaching elementary school mathematics well demands what Ma (1999) has called a "profound understanding of fundamental mathematics," that is, knowledge of mathematics that is rooted in conceptual understanding and in the modes of inquiry of the discipline (Ball, 1990). This kind of understanding of mathematics does not easily develop behind close doors; rather, as Ma's teachers reported, it flourishes in collaborative conversations and inquiry with colleagues. The study group model we have developed is one way in which elementary teachers can engage in the study of mathematics and mathematics teaching while keeping the conversations situated in the participants' own teaching practice. In the SATRR model, there is a generative back and forth between the study group conversations and each teacher's teaching; the study group conversations about mathematics problems shape the participants' teaching experiments and their experiences with this mathematics teaching in turn inform subsequent study group conversations.

Although we have reported here on the multiple learning opportunities that the SATRR model for mathematics study groups creates, we have also alluded to some of the challenges the model poses for both participants and facilitators. Our closing thoughts highlight some of these challenges. First, we want to emphasize that we, like others who have organized teacher study groups as a form of professional development (e.g., Arbaugh, 2004; Grossman et al., 2001), have needed to negotiate multiple dilemmas in initiating and sustaining the groups. For example, in each school, we needed to decide who should participate. If participation is mandated, the facilitator is likely to encounter major obstacles in engaging the teachers in collective inquiry of mathematics and teaching. Yet teachers who join voluntary groups may be those in least need of professional development.

Second, we call attention to the importance of selecting carefully the mathematics problems to be used with the group. Although study group members may ask facilitators to find problems located in the particular topics they are teaching (subtraction with regrouping, for example), problems chosen to provide short term help to one or two teachers rarely engage the entire group effectively. Problems should be chosen because they challenge both adults and elementary students. We have found that good study group tasks – ones that lead to productive mathematical conversations in the teacher study group and that also adapt well for use in K-5 classrooms – have several characteristics.

Problems whose solutions involve noticing patterns (like the trikes and bikes problem) are good candidates for SATRR groups. Such problems usually engaged the teachers in our groups and could be readily scaled for children at different grade levels; moreover, the teachers enjoyed observing and discussing what their students did with such problems. Problems that highlight challenges to mathematical reasoning and communication led to interesting mathematical insights in the study group and generated provocative mathematical and pedagogical discussions (as did the Pizza Problem, NCTM, 2000 and the Fair Game problem, NCTM, 1991, both are featured in Crespo, in press).

Furthermore, a good problem must challenge all the participants. If some teachers were able to use an algorithm to solve the problem quickly, the conversation in the study group was easily aborted and invisible hierarchies within the group were reinforced. However, facilitators must walk a tight rope in relation to the difficulty of the task. If the mathematics problem fails to engage the teachers in genuine mathematical inquiry (if it is too easy) then the kind of explorations that the teachers will design for their students will fall short in the same way. Yet if the problem poses too great a challenge — if at the end of the meeting only a few of the teachers feel that they really understand the mathematics of the problem – most teachers will be unable to use it to create a mathematically-worthwhile lesson for their students.

Finally, we draw attention to the challenge of facilitating teacher groups. Helping the group become a community and not simply a gathering of teachers is complex and time-consuming work, as Grossman and her colleagues (2001) have shown. Participants may understand better than facilitators, who are the outsiders to the school culture, that sharing stories of practice with an unselected group of colleagues can be dangerous. The facilitator must respect the fact that she is less vulnerable than the teachers while continuing to support those who volunteer ideas and stories and to encourage those who hold back.

The facilitator must walk a similar tightrope concerning the group's work of adapting problems for younger children. The teachers will understand the pedagogical and mathematical potential of an unfamiliar math problem much better *after* they teach it than they do before, but they must adapt it *before* they teach it. Therefore, inevitably, some teachers will propose adaptations that turn the problem into a computation exercise, removing the mathematics along with much of the challenge (see Stein, Smith, Henningsen, & Silver, 2000, for an analysis of ways that teachers transform the cognitive demand of math problems). Part of the facilitator's struggle is to figure out how much to 'police' the adaptation that the teachers propose to take to their classrooms and how much to redirect what one might consider unproductive conversations.

The practice of facilitating teacher groups is largely undocumented. Only recently have teacher study groups become an area of research interest and attention. There is little in print that can help future teacher group facilitators to prepare for this practice. Considering our own continued quest to improve our practice as facilitators of teacher groups, we do not find it surprising that much of what we have learned has come from our own personal and collaborative inquiries. We have drawn from our experience facilitating discourse within our own mathematics and teacher education classrooms and from reading the research literature on teacher learning communities and classroom discourse practices. Our experience suggests that participating in a community of facilitators that comes together on a regular basis to share and analyze their practice helps group leaders as much as it helps teachers.

As this enumeration of dilemmas and difficulties facing those who would organize and facilitate SATRR groups demonstrates, we share the concerns of the many math educators and scholars who have cautioned against romanticizing the teacher study group as a means of improving teaching practice. We note, however, that Ma's careful study of Chinese and U. S. teachers has shown that teachers who begin their careers with only a superficial understanding of school mathematics can develop profound understanding of elementary mathematics while teaching in a school whose culture encourages collegial conversation about mathematics and teaching.

Although we cannot reproduce Chinese faculty culture in U. S. public elementary schools, the study group model is one way in which U. S. elementary teachers can engage in the study of mathematics and mathematics teaching while keeping the conversations directly situated in the participants' own teaching practice. This model respects the fact that elementary school teachers care deeply about the learning of the children who return to their classrooms every morning. These teachers will put faith and effort into any intervention that engages these children intellectually and emotionally and makes the intelligence and intellectual creativity of 'unsuccessful' students newly visible. As the funding for the project ended, teachers in our SATRR groups asked facilitators to continue the group meetings. To us this is an important indicator of the potential that teacher groups offer to improving mathematics teaching in classrooms. In spite of the challenges and time commitment, teachers come to value the experience and seek out ways to continue to participate in this kind of professional development.

References

Arbaugh, F. (2003). Study groups as a form of professional development for secondary mathematics teachers. *Journal of Mathematics Teacher Education, 6*, 139-163.

Ball, D. L. (1990). The mathematical understandings that preservice teachers bring to teacher education. *Elementary School Journal, 90*(4), 449-466.

Ball, D., & Cohen, D. (1999). Developing practice, developing practitioners: Toward a practice-based theory of professional education. In L. Darling-Hammond and G. Sykes (Eds.), *Teaching as the learning profession: Handbook of policy and practice* (pp. 3-53). San Francisco: Jossey-Bass.

Blythe, T., Allen, D., & Powell, B. S. (1999). *Looking together at student work.* New York: Teachers College Press.

Burns, M., & Tank, B. (1988). *A collection of math lessons: From grades 1 through 3.* New Rochelle, NY: Cuisenaire Company.

Carroll, D., & Thompson, J. (2002). *Study group leadership: Learning in and through practice.* Paper presented at the annual meeting of the American Educational Research Association, April 2002, New Orleans.

Chazan, D., & Ball, D. (1999). Beyond being told not to tell. *For the Learning of Mathematics, 19*(2), 2-10.

Crespo, S. (in press) Elementary teachers talk in mathematics study groups. *Educational Studies in Mathematics.*

Crespo, S. (2002). Teacher learning in mathematics teacher study groups. In D. S. Mewborn, P. Sztajn, D. Y. White, H. G. Bryant, & K. Nooney (Eds), *Proceedings of the North American Chapter of the International Group for the Psychology of Mathematics Education, 24,* (vol 3, pp. 1439-1450). Columbus, OH: ERIC Clearinghouse for Science, Mathematics, and Environmental Education.

Crespo, S., & Featherstone, H. (2003; 2002; 2001). *Communities of practice to improve mathematics and mathematics teaching: Years 1-3.* Technical Reports to the Lucent Technologies Foundation. East Lansing, MI: Michigan State University.

Crespo, S. M., Kyriakides, A. O., & McGee, S. (2005). Nothing "basic" about basic facts: Exploring addition facts with fourth graders. *Teaching Children Mathematics, 12*(2), 60-67.

Fernández, C., & Yoshida, M. (2004). *Lesson Study: A Japanese approach to improving mathematics teaching and learning.* Mahwah, NJ: Lawrence Erlbaum Associates.

Grossman, P., Wineburg, S., & Woolworth, S. (2001). Toward a theory of teacher community. *Teachers College Record, 103*, 942-1012.

Kazemi, E., & Franke, M. (2004) Teacher learning in mathematics: Using student work to promote collective inquiry. *Journal of Mathematics Teacher Education, 7*, 203-235.

Little, J. W. (1990). The persistence of privacy: Autonomy and initiative in teachers' professional relations. *Teachers College Record, 91*, 509-536.

Lord, B. (1994). Teachers' professional development: Critical colleagueship and the role of professional communities. In N. K. Cobb, (Ed). *The future of education perspectives on national standards in America* (pp. 175-204). New York: College Entrance Examination Board.

Lortie, D. (1975). *Schoolteacher: A sociological study*. Chicago: University of Chicago Press.

Ma, L. (1999). *Knowing and teaching elementary mathematics: Teachers' understanding of fundamental mathematics in China and the United States*. Mahwah, NJ: Lawrence Erlbaum.

McLaughlin M., & Talbert, J. (2001). *Professional communities and the work of high school teaching*. Chicago: The University of Chicago Press.

National Council of Teachers of Mathematics. (2000). *Principles and standards for school mathematics*. Reston, VA: Author.

National Council of Teachers of Mathematics (1991). *Professional standards for teaching mathematics*. Reston, VA: Author.

Pfeiffer, L. C., & Featherstone, H. J. (1995). *Toto I don't think we're in Kansas anymore: Entering the land of public disagreements in learning to teach.* Research Report 97-3, National Center for Research on Teacher Learning: Michigan State University, East Lansing. http://ncrtl.msu.edu/http/rreports/html/pdf/Rr9703.pdf

Putnam, R., & Borko, H. (2000). What do new views and knowledge and thinking have to say about research on teacher learning? *Educational Researcher, 29*(1), 4-15.

Stein, M., Smith, M. S., Henningsen, M., & Silver, E. (2000). *Implementing standards-based mathematics instruction: A casebook for professional development.* New York: Teachers College Press.

White, D., Sztajn, P., Hackenberg, A., & Snider, M. A. (2004) Building a mathematics education community that facilitates teacher sharing in an urban elementary school. In D. McDougall and J. Ross (Eds.), *Proceedings of the twenty-sixth annual meeting of the North American Chapter of the International Group for the Psychology of Mathematics Education* (volume 3, pp. 977-983). Toronto, ON: OISE/UT.

Wilson, S. M., & Berne, J. (1999). Teacher learning and acquisition of professional knowledge: An examination of contemporary professional development. In A. Iran-Nejad & P. D. Pearson (Eds.), *Review of Research in Education, 24,* 173-209. Washington, D.C.: American Educational Research Association.

---

[i] This is a variation of the well-known "cows and chickens" problem (given number of feet, how many of each animal). Ours is an adapted version of the tricycles and bicycles problem-solving lesson described in Burns & Tank (1988). *A Collection of Math Lessons: From Grades 1 Through 3.* New Rochelle, NY: Cuisenaire Company.

Sandra Crespo is an associate professor of teacher education at Michigan State University. Her research focuses primarily on preservice elementary teachers and their development as learners of mathematics and mathematics teaching. She became interested in teacher groups as contexts for teacher learning when she began to work closely with teacher mentors. She has collaborated with Featherstone and other colleagues to explore school-based teacher groups to support the mentor teachers' professional development as teachers of mathematics and mentors of preservice teachers who visit their classrooms.

Helen Featherstone is an associate professor of teacher education at Michigan State University. She is particularly interested in teachers' efforts to change their practices of mathematics teaching. Her research focuses on the learning of teachers, prospective teachers, and teacher educators. She is working on two books, one on exemplary teacher education programs and one on the teaching and learning of an elementary teacher and a teacher educator during a five week collaborative effort to teach third graders about operations of integers in the teachers' urban classroom.

Silver, E. A., Mills, V., Castro, A., and Ghousseini, H.
AMTE Monograph 3
The Work of Mathematics Teacher Educators
©2006, pp. 117-132

8

Blending Elements of Lesson Study with Case Analysis and Discussion: A Promising Professional Development Synergy[1]

Edward A. Silver
University of Michigan

Valerie Mills
Oakland (MI) Schools

Alison Castro
University of Illinois-Chicago

Hala Ghousseini
University of Michigan

Although there are many approaches to professional development for mathematics teachers, no single approach is perfect. Rather than choosing only one from a menu of options, it should be possible to blend different approaches to use the strengths of each in complementary fashion. We illustrate this idea by drawing on examples from the BIFOCAL project, in which we integrate two popular approaches to mathematics teacher professional development: lesson study and case analysis/discussion. We show how the synchrony of the two approaches creates powerful opportunities for teachers to inquire into mathematics instruction in ways that influence their teaching practice.

Public and professional discourse regarding mathematics education often reflects differing views about curricular goals, curriculum materials, and teaching methods. Yet, there appears to be nearly universal agreement on one point: Substantial improvement in

---

[1] This article is based upon work supported in part by the National Science Foundation under Grant No. 0119790 to the Center for Proficiency in Teaching Mathematics and in part by the Michigan State University Mathematics Education Endowment Fund for support of the BIFOCAL project. Any opinions, findings, and conclusions or recommendations expressed in this material are those of the authors and do not necessarily reflect the views of the National Science Foundation, the Center, or the university.

students' mathematics achievement is unlikely to occur without serious attention to both the initial preparation and ongoing professional development of teachers of mathematics (Cohen & Hill, 2002; Darling-Hammond & Sykes, 1999). The improvement of the professional practice of teachers is seen as integral to the effort to improve K-12 education. This view is consistent among strong advocates for, as well as staunch opponents of, "standards-based" reform in education. There is a growing recognition that, with increasing understanding of the complexity of good mathematics teaching, opportunities for teachers to improve their practice continually are critical. This point was made succinctly in *Adding it Up*, the National Research Council report on mathematics education in the United States: "Professional development beyond initial preparation is critical for developing proficiency in teaching mathematics" (National Research Council, 2001, p. 399).

In an effort to meet the growing demand for high quality professional development for teachers of mathematics, researchers and educators have developed and made available a variety of approaches intended to help teachers improve their practice. Nevertheless, despite a plethora of good ideas and materials, decisions regarding the design of professional development experiences for mathematics teachers have an unfortunate tendency to devolve into a process similar to answering a "multiple-choice" test item. According to this view, it is necessary to choose *only one* from a menu of attractive and useful professional development approaches. That is, video cases compete with narrative cases, and each competes with lesson study and with curriculum-based approaches. The professional developer's task, then, is to choose from this array of options the one that is optimal. In contrast, we argue that these different approaches have strengths and limitations; by carefully and intentionally blending approaches, we can use the strengths of one to complement the limitations of the other.

In this article, we amplify and illustrate this argument as it applies to two popular approaches to mathematics teacher professional development: lesson study and case analysis/discussion. To do this, we draw on experience in and data from the BIFOCAL (Beyond Implementation: Focusing on Challenge and Learning) Project, in which we use particular versions of these two professional development strategies in our work with middle school mathematics teachers.[i]

Conceptualizing a Professional Development Synergy

There is a basis for optimism about blending lesson study with case analysis and discussion. Each approach has strengths and limitations, and the strengths of each nicely address the weaknesses of the other. In particular, case analysis and discussion can be used to build teachers' proficiency with the following intellectual practices and professional dispositions that are needed for successful use of lesson study:

- Treat classroom instruction as an object of inquiry in discussions with colleagues;
- Adopt an analytic stance toward teaching in general;
- Learn to make claims based on evidence rather than opinion;
- Attend to general instructional goals and issues; and
- Consider a classroom lesson as a unit of *analysis*.

Similarly, lesson study may complement and enhance the effects of case analysis and discussion by assisting teachers to become more proficient in the following intellectual practices and professional dispositions that are needed for instructional improvement:

- De-privatize classroom instruction within a professional community so that others can learn from it;
- Adopt an analytic stance toward one's own teaching;
- Commit to the steady improvement of teaching;
- Analyze general instructional issues in relation to one's teaching; and
- Consider a classroom lesson as a unit of *improvement*.

Our conceptualization of the complementary strengths of lesson study and case analysis suggests that they should complement each other and that these two approaches can be integrated to achieve a powerful synergy. In the remainder of this paper, we describe a project in which this has been done with particular versions of the two approaches. In addition, we offer an example of how blending the two approaches created opportunities for teachers to make progress on an instructional issue of concern to them.[ii]

Overview of the BIFOCAL Project

BIFOCAL is aimed at solving a problem that might be called the *curriculum implementation plateau*. When innovative curriculum materials are adopted, some professional development support is typically offered to familiarize teachers with the new materials; however, the quality, amount, and duration of support is usually insufficient to ensure effective and masterful use of the materials to promote student learning. Consequently, a few years after initial implementation, teachers may become comfortable with certain aspects of the new curriculum materials (e.g., lesson structure, topical sequence), but they may not yet have gained proficiency in using the curriculum materials for maximum effectiveness. These teachers might be said to have reached a curriculum implementation plateau. In this project, we work with users of *Connected Mathematics*, a standards-based, middle school mathematics curriculum, to address this problem.[iii]

Project Foundations

Problems are central to mathematics teaching and learning and constitute the basis for intellectual activity in the classroom (Lampert, 2001; Stein, Smith, Henningsen, & Silver, 2000). The BIFOCAL Project seeks to enhance teachers' mathematical knowledge, their attention to and use of student thinking, and their proficiency with a repertoire of instructional strategies to facilitate effective use of innovative curriculum materials to promote student learning through their productive engagement with mathematical tasks. Helping teachers learn to support student engagement with complex intellectual activity is particularly important because such instruction rarely is found in U.S. mathematics classrooms according to large-scale studies of classroom practice in middle grades mathematics (Stigler & Hiebert, 1999).

BIFOCAL builds on a foundation of prior work in the QUASAR and COMET projects, particularly the effective use of cognitively demanding tasks in the classroom (e.g., Silver & Stein, 1996; Stein, Grover & Henningsen, 1996; Stein, Smith, Henningsen & Silver, 2000). Central to our work is the Mathematical Task Framework (MTF) developed by Stein and colleagues (1996; 2000).[iv] The MTF portrays a mathematical task as passing through phases, from the task as found in curricular materials, to the task as it is set up by the teacher in the classroom, to the task as enacted by students and their teacher interacting with each other and with the task during the lesson. The MTF underscores the important role that mathematical tasks play in influencing students' learning opportunities. Moreover, the MTF points to the centrality of the work that teachers do with and around mathematical tasks. Teachers' decisions and actions influence the nature and extent of student engagement with challenging tasks, and ultimately affect students' opportunities to learn from their work on such tasks.

The MTF provides a framework within which to consider some of the challenges that teachers' face when using complex tasks in the mathematics classroom. In particular, teachers must decide "...what aspects of a task to highlight, how to organize and orchestrate the work of the students, what questions to ask to challenge those with varied levels of expertise, and how to support students without taking over the process of thinking for them and thus eliminating the challenge"

(National Council of Teachers of Mathematics, 2000, p. 19). To do this well, teachers need a fluent repertoire of instructional routines, skill in designing instruction, knowledge of how students think and learn about mathematics, and a profound knowledge of mathematics to create learning opportunities for students. Teachers need to resist the persistent urge to tell students precisely what to do, thereby removing the opportunity for thoughtful engagement; rather, they need to respond to student queries and requests for information in ways that support student thinking rather than replace it.

The design of the project has been influenced by prior work on effective professional development, particularly the need for long-term engagement with issues closely related to the work of teaching (e.g., Cohen & Hill, 2001; Loucks Hoursley et al., 1998). BIFOCAL employs a *practice-based* approach (Ball & Cohen, 1999; Smith, 2001), grounding the professional development explicitly in critical aspects of teaching practice, such as lesson preparation and the classroom enactment of lessons involving complex mathematical tasks in ways that enhance students' opportunities to learn.

Project Operation

The project has unfolded in phases. In phase 1 (July 2002 to June 2004), we identified local school districts that had been using S*tandards*-based middle-grades curriculum materials for at least 3 years and solicited indications of potential interest in participating in the project. Four small school districts with a total of five middle schools (grades 6-8), each of which had been using *Connected Mathematics* for several years, agreed to participate, after which we engaged a few teachers from each school in a year-long sequence of professional development sessions. In phase 2 (July 2004-June 2005), the project was expanded in two ways – additional participants were added from the participating schools and districts, and school-based professional development sessions (organized and facilitated by the participants from phase 1) were interwoven with large group sessions during the school year. Given page limitations in this article, we focus only on phase 1.

Each school sent a team composed of two or three middle school teachers; each school's principal also participated in some sessions. Professional development began with a day-long session in May 2003 and a two-day residential experience in June 2003. During the 2003-04 academic year, the participants' group met roughly once per month in a series of eight day-long seminar sessions that attended to their personal professional development as mathematics teachers and also, to some extent, to the development of skills as teacher leaders. We discuss here only the first emphasis.

During the year, the twelve teachers met regularly in sessions designed to assist them to improve their teaching practice by engaging in two main sets of activities: (1) case analysis and discussion and (2) modified lesson study.[v] We describe each of these activities briefly, after which we illustrate the interplay between the two modes of professional development in the work that teachers do in the project.

Case Analysis and Discussion

A critical part of the case analysis and discussion work in BIFOCAL involved engaging the participants with mathematical ideas embedded in the case. Before examining a case, participants solved a mathematical task drawn from the case, considered several possible solution strategies, and identified mathematical goals that might be advanced if this task were used with students.

The cases we used were of several types: (a) narrative mathematics instructional cases found in *Implementing Standards-based Mathematics Instruction: A Casebook for Professional Development* (Stein, Smith, Henningsen & Silver, 2000) and pre-published versions of cases available from the COMET Project[vi]; (b) publicly released video clips from the Third International Mathematics and Science Study (TIMSS) video study that depict teaching in classrooms where complex mathematical tasks were being used; and (c) collections of samples of student work on intellectually demanding mathematics tasks in the *Connected Mathematics* curriculum. Each case was selected carefully to stimulate reflection, analysis, and inquiry aimed at one or more of the challenges faced by teachers working with cognitively challenging tasks in the middle grades mathematics classroom. The narrative and video cases portray the relationships among mathematics, pedagogy, and student learning in the classroom, rather than fragmenting teaching strategies, student thinking, and mathematical ideas. Thus, as a reader analyzes a case, he or she can examine the components of teaching individually or as a complex interaction.

Lesson Study

During the sessions, teachers also completed several cycles of a modified lesson study process – selecting a target lesson, planning collaboratively, teaching the target lesson, reflecting on their instructional moves in relation to evidence of students' thinking and understanding, and analyzing their lessons in discussions with colleagues. We used the *Thinking through a Lesson Protocol* (Smith & Bill, 2004)[vii] to focus and support teachers in their collaborative planning and their lesson reflection and analysis. This provided a common framework for lesson planning and reflection/analysis.

Teachers generally began the modified lesson study portion of the session by discussing the enactment of the previous months' lesson, especially attending to issues of cognitive demand in the task enactment. The teachers worked in small groups to identify a lesson from their curriculum that they would be teaching in the near future and in which the issues identified in the morning's case discussion would be likely to surface. Prompted by questions prepared by the facilitator to direct their attention toward the session's instructional focus, the teachers collaboratively planned the lesson they agreed to implement in their individual classrooms. So, throughout the day, teachers proposed and explored ideas related to instruction that arose from the case discussion and then analyzed and tested the applicability of these ideas in their lesson planning. In a subsequent session, the teachers analyzed the efficacy of the ideas as they were represented in the lessons they taught and revisited the ideas in ensuing cases, thereby refining and extending their understanding of the ideas.

Illustrating the Professional Development Synergy: A Look Inside BIFOCAL

This example draws from project sessions occurring over several months during phase 1 of the project (beginning with the first session in May 2003 and culminating about one year later in April 2004). We focus on teachers' consideration of the role of multiple solution strategies in classroom work with complex mathematics problems.

May 2003. Participants began the first session by working on the mathematics task that was the focus of the narrative case they would read. The task required finding the perimeter of a "train" formed by 3 adjacent hexagonal "cars," then the perimeter for a train with 4 cars, then for 10 cars, and finally, *n* cars.[viii] After working individually, participants presented different solution strategies to the group. Participants were also asked how they thought their students would approach and solve this task. All of this was intended to engage the participants in mathematical work and to set the stage for their reading, analysis, and discussion of the case.

During the discussion of participant solutions and anticipated student work, one issue that arose was the value of considering multiple solutions: Do all students benefit from considering more than one solution? After a brief exchange of differing views, the discussion facilitator decided to leave the issue unresolved, knowing that it was likely to surface again in analysis and discussion of the narrative case.

Participants read the narrative case, considering how two teachers' actions and reactions appeared to support or inhibit students' opportunities to learn. As expected, the issue of multiple solution strategies arose again in relation to ways that the teachers in the case did or did not invite consideration of different student solutions in their lessons. When asked how and why consideration of multiple strategies might affect students' opportunities to learn, one participant commented, "...instead of focusing on just one student, you help everyone feel comfortable to give their opinion, or share their strategy or their way of how they looked at it..." Although there appeared to be agreement that multiple solutions were desirable, a few participants voiced concerns about possibly confusing students with multiple solutions and appeared to be reluctant to consider multiple solution strategies when teaching a lesson for the first time or if students might not be familiar with the ideas involved in a solution.

January and February 2004. Between May 2003 and January 2004, participants attended several sessions in which they considered another issue that emerged in the May 2003 session – scaffolding student thinking. We selected cases intended to draw attention both to anticipating student thinking in preparing for a lesson and to scaffolding student thinking as a lesson unfolds. Each case used during these sessions portrayed an instructional episode in which multiple solution strategies were displayed. Although handling multiple student solutions in a classroom lesson was not the primary focus of these case discussions, it is clear that some teachers attended to this issue as they thought about what might be applicable to their own teaching. In post-session reflections from Fall 2003 sessions, some participants indicated that they had been actively thinking about whether and how to incorporate the use of multiple solution strategies in their lessons.[ix]

For the January 2004 session, we selected a student work sample case based on The Orange Juice (OJ) Task, a proportionality task that appears in the *Connected Mathematics Project* (CMP) curriculum materials in the *Comparing and Scaling* unit (Lappan, Fey, Fitzgerald, Friel, & Phillips, 2004). The OJ Task gives three different mixtures of orange flavoring and water and requires students to determine the mixture with the strongest orange taste. The case we used with teachers included the task and six samples of student work, each exhibiting a different solution approach.[x]

This case was relevant to several of our professional development goals at this time. It provided an opportunity to focus on the use of proportions and proportional reasoning, which had surfaced in an earlier session as an area of possible weakness for at least some of the teachers. We also wanted to provoke participants to think deeply about the interplay between student thinking and the mathematical goals of a lesson. This case allowed us to return to the issue of teachers working with multiple student solution strategies in a lesson.

Typically, teachers worked on the mathematics task in the case, first solving it, and then thinking about different ways students would likely work on it. They discussed different solution strategies, with the facilitator drawing attention to their use of part-to-part and part-to-whole comparisons within proposed solutions (which was an important difference among the six student solutions that they would soon consider). After solving and discussing the problem and identifying the mathematical goals that might be addressed by the task, teachers turned their attention to the case, which was the set of six student work samples. They looked at each solution, determined the solution approach that each represented, and grouped the six samples into subsets based on similarities in student thinking evident in the solutions. They also identified the solutions that they thought were important to display and discuss with their students, and to think about how they would do this.

The project facilitator asked the participants about the value for students in seeing all of these solution approaches during the discussion of the student work samples. Two teachers commented that being exposed to different strategies would provide students with "alternative thinking," which they thought would likely prepare students to be willing to try new problems. However, one teacher voiced her concern and confusion: "...if we're doing this [OJ Task] and we're trying to show alternate strategies...with the time crunch, do you try to make sure that the class as a whole understands each of the strategies? Or do you make sure they understand one well?" She continued by saying, "I think that sometimes I'm scared to put, you know two different strategies up there because they were barely able to get one and then you start throwing an alternate up there..."

In the subsequent discussion and in post-session reflections, it appeared that several teachers were comfortable in handling multiple solution strategies in their classrooms, but others were still struggling with whether and how to present students with multiple solutions to a task. In her reflection, one teacher wrote, "...I am still trying to figure out what and when to share student work." Yet another teacher mused about how to handle solutions that she might have wanted available but that were not generated by students: "...do we need to share strategies that are not brought up?" We decided to continue work on this issue in the next session given the persistence of teacher concerns related to this issue.

For the February session we selected a narrative case that would provide another opportunity to consider the issue of presenting multiple solution strategies during instruction. This case features a lesson in which the teacher made careful use of a variety of student solutions to advance her mathematical goals. The analysis and discussion of this case afforded yet another opportunity to bring to the surface the concerns of some teachers about whether seeing different approaches would confuse students. It offered an example of a teacher skillfully orchestrating a lesson in which students considered multiple solution approaches. In their collaborative lesson planning, teachers were asked to select a CMP lesson in which they expected the issue of working with multiple student solution strategies to arise; they were given a set of structured questions to support their planning and focus their attention on cognitive demands and multiple student solutions.

March 2004. The goals of this session were to resume consideration of instructional moves and decisions, with particular attention to the use of multiple solution strategies, and facility with scaffolding strategies and mathematical questions that might allow teachers to maintain high cognitive demand as students worked on a task. During the modified lesson study portion of the March session, teachers met in grade-level groups to debrief the enactment of the lessons selected and planned at the February session.

One teacher, Natalie, explicitly tackled the issue of multiple solution strategies in her lesson, drawing on ideas first made visible to her in the January session in the case discussion of the OJ task. When describing her lesson, Natalie shared posters of her students' work while she explained to the group how the lesson unfolded. Several important points surfaced:

- Natalie noted that she wanted to try a teaching technique in this lesson that built on something she saw the facilitator do in the January session when the multiple student solutions for the OJ Task were discussed, namely, the intentional sequencing of solution strategies to facilitate comparing and contrasting them.
- In planning for this lesson, Natalie carefully selected the subset of questions she would ask students to discuss as a whole group, following their initial paired problem solving. She selected questions on the basis of the lesson's mathematical goals.
- Natalie walked around the room and noted the different solution approaches that students were taking while they were working on the task.
- She "ranked" students' solution approaches according to her judgment of their sophistication. For example, using a guess-and-check method received a 1; solutions with more complex representations and strategies were ranked as 2, 3, or 4.

- Natalie facilitated the discussion of solutions by inviting students to present solutions based on their rank ordering.
- She began by having students present different solutions each ranked 1, after which she orchestrated a discussion of similarities and differences between solutions.
- Natalie then had different solutions ranked 2, 3 and 4 presented, after which there was also a discussion of relationships among solutions, as well as their relative strengths and limitations.

Natalie's presentation generated a lively interchange among the participants during which many pragmatic issues were considered. In their post-session reflections, all but one of the teachers stated they intended to integrate multiple solution strategies into an upcoming lesson, and several made specific reference to Natalie's presentation: T1 - "I want to implement sharing students' work in the way shared by Natalie." T2 - "I want to use Natalie's idea of putting numbers on student work for order of presentation to class…" These comments suggest that Natalie's approach to operationalizing some instructional routines related to the sharing of multiple solution strategies was seen as useful to some teachers. However, the impacts noted in the teacher comments were not all related to the specifics of Natalie's presentation; others reflected more cumulative development related to this issue. For example, one teacher wrote, "I am learning more and more that having kids present their work can have such an impact on their learning." Moreover, in end-of-year interviews conducted with participants, more than half the teachers mentioned the issue of multiple solution strategies as one on which they felt they had made progress.

Conclusion

At the beginning of this paper, we argued that curriculum-based professional development is probably *necessary* but perhaps not *sufficient* to support teachers' continued growth and development. In BIFOCAL, we built upon a foundation of teacher access to and familiarity with a curriculum that offers rich, complex mathematical tasks. We blended elements of lesson study and case analysis/discussion to assist teachers in deepening their understanding of challenges they face in using such a curriculum and increasing their proficiency in meeting the challenges so they can support student engagement in complex mathematical activity in the classroom.

Our experiences in the BIFOCAL project illuminate ways in which the synchrony of lesson study and case analysis/discussion creates powerful opportunities for teachers to examine the practice of mathematics teaching and to learn from this examination in ways that influence their own teaching practice. Using narrative, video, or student work sample cases, teachers are situated safely outside of their own teaching and have opportunities to examine critically the practice of teaching in ways that are "not about them" but that are nevertheless applicable to their work. In particular, cases create a space for teachers to read, analyze, and discuss a variety of issues related to instructional practices and student thinking. In the example discussed above, we saw how issues that initially arose in connection to the "Catherine Evans & David Young" case (Smith, Silver & Stein, 2005a) were carried throughout a year-long series of professional development sessions. We also saw that the modified lesson study experience afforded opportunities for teachers to work on and with the issues and ideas generated in case discussions as they worked with colleagues to plan, enact, and analyze lessons taught in their own classrooms.

In this chapter we have illustrated how participants' work with cases – including the time they spend considering mathematical ideas prior to reading/viewing, analyzing, and discussing a case – can generate issues and insights that carry into collaborative lesson planning, which in turn generates renewed consideration of these issues/insights or the generation of new ones that can be taken up in other cases, and so on. Through this cycling process built around the synchrony of two professional development approaches, important and useful ideas, issues, and insights can "travel" in ways that influence teachers' instructional practice.

References

Ball, D. L., & Cohen, D. K. (1999). Developing practice, developing practitioners: Toward a practice-based theory of professional education. In G. Sykes and L. Darling-Hammond (Eds.), *Teaching as the learning profession: Handbook of policy and practice* (pp. 3-32). San Francisco: Jossey Bass.

Cohen, D., & Hill, H. (2001). *Learning policy.* New Haven, CT: Yale University Press.

Darling-Hammond, L., & Sykes, G. (1999). *Teaching as the learning profession: Handbook of policy and practice.* San Francisco: Jossey-Bass Publishers.

Lampert, M. (2001). *Teaching problems and the problems of teaching.* New Haven, CT: Yale University Press.

Lappan, G., Fey, J., Fitzgerald, W., Friel, S., & Phillips, E. (2004). *Comparing and scaling* (one unit in *Connected Mathematics*, Grade 7). Glenview, IL: Prentice-Hall.

Loucks-Horsley, S., Hewson, P. W., Love, N., & Stiles, K. E. (1998). *Designing professional development for teachers of science and mathematics*. Thousand Oaks, CA: Corwin Press.

National Council of Teachers of Mathematics. (2000). *Principles and standards for school mathematics*. Reston, VA: Author.

National Research Council. (2001). *Adding it up: Helping children learn mathematics.* J. Kilpatrick, J. Swafford, B. Findell (Eds). Mathematics Learning Study Committee, Center for Education, Division of Behavioral and Social Sciences and Education. Washington, DC: National Academy Press.

Silver, E. A., & Stein, M. K. (1996). The QUASAR project: The "revolution of the possible" in mathematics instructional reform in urban middle schools. *Urban Education*, *30*, 476–521.

Smith, M. S. (2001). *Practice-based professional development for teachers of mathematics*. Reston, VA: National Council of Teachers of Mathematics.

Smith, M. S., & Bill, V. (2004, January). Thinking through a lesson: Collaborative lesson planning as a means for improving the quality of teaching. Presentation at the annual meeting of the Association of Mathematics Teacher Educators, San Diego, CA.

Smith, M. S., Silver, E. A., & Stein, M. K. (2005a). *Improving instruction in algebra: Using cases to transform mathematics teaching and learning, Volume 2*. New York: Teachers College Press.

Smith, M. S., Silver, E. A., & Stein, M. K. (2005b). *Improving instruction in rational numbers and proportionality: Using cases to transform mathematics teaching and learning, Volume 1*. New York: Teachers College Press.

Stein, M. K., Grover, B. W., & Henningsen, M. A. (1996). Building student capacity for mathematical thinking and reasoning: An analysis of mathematical tasks used in reform classrooms. *American Educational Research Journal, 33*, 455-488.

Stein, M. K., Smith, M. S., Henningsen, M. A., & Silver, E. A. (2000). *Implementing standards-based mathematics instruction: A casebook for professional development*. New York: Teachers College Press.

Stigler, J., & Hiebert, J. (1999). *The teaching gap*. New York: The Free Press.

---

[i] The BIFOCAL project is one of several activities sponsored by the Center for Proficiency in Teaching Mathematics (CPTM), which is a collaborative venture of the University of Michigan and the University of Georgia. BIFOCAL is also supported by a grant from the Michigan State University Mathematics Education Endowment Fund.

[ii] There are different ways to use case analysis and discussion as a professional development approach, and different variations in the way that lesson study is used. We think our general argument regarding the synergy of the two approaches applies fairly well across all variations. Nevertheless, we offer examples that depend heavily on the particular way in which we use case analysis and discussion and the elements of lesson study that we adopt in BIFOCAL.

[iii] Information about the Connected Mathematics curriculum materials is available at http://www.math.msu.edu/cmp/Overview/Glance.htm.

[iv] A more complete explanation of the Mathematics Tasks Framework is provided by Stein et al. (2000); see especially pages 1-32.

[v] We use the term *modified* lesson study to signal that our process is not identical to Japanese lesson study as it has been described by experts on this form of professional development. Nevertheless, we think it is appropriate to call this a modified form of lesson study because it possesses many similar features, such as shared goal setting, collaborative lesson planning, and a common analytic framework for reflecting on the lesson.

[vi] COMET cases are now available in published form (Smith, Silver & Stein, 2005a, 2005b).

[vii] The *Thinking through a Lesson Protocol* is available at http://www.cometproject.com/alg/Assignment%202.pdf.

[viii] The Hexagon-Pattern Task used in this session was identical to that now found in the published version of the COMET cases. This is the Opening Activity of "Examining Linear Growth Patterns: The Case of Catherine Evans and David Young" (Smith, Silver, & Stein, 2005a, p. 9).

[ix] At the end of each monthly session, teachers wrote an End-of-Day reflection stating whether that session had caused them to think about any issues related to their own teaching and student thinking.

[x] The task and student work samples we used in our session are available from the Connected Mathematics Project at http://www.math.msu.edu/cmp/RREvaluation/StudentWork.htm.

Edward Silver is the William A. Brownell Collegiate Professor in Education and the Associate Dean for Academic Affairs in the School of Education at the University of Michigan. He teaches and advises graduate students in mathematics education, conducts research on the teaching and learning of mathematics, and engages in a variety of professional service activities. He is co-PI of the NSF-funded Center for Proficiency in Teaching Mathematics (http://www.cptm.us). He also co-directs BIFOCAL, in which materials from the COMET project (http://www.cometproject.com) are being used to support the learning of mathematics teachers in several local Michigan school districts.

Valerie Mills is Mathematics Education Consultant and Unit Supervisor for Oakland Schools, a Michigan intermediate agency serving 28 districts and 260,000 students. Previously she served as high school mathematics teacher, mathematics coordinator and curriculum director. She is currently co-director of BIFOCAL, PI for the Mathematics Education Resource Center, a Mathematics/Science Partnership project with 15 urban schools, and senior author for Lenses on Learning Secondary, an NSF-funded professional development series for secondary administrative and teacher-leader mathematics teams.

Alison Castro is an Assistant Professor of Mathematics Education at the University of Illinois at Chicago. Her work focuses on how preservice and inservice teachers use mathematics curriculum materials to create opportunities for students to learn mathematics, with particular attention to teachers' mathematical knowledge, lesson planning, and instructional enactment strategies. She has been a member of the BIFOCAL Project research team since its inception. During her time as a doctoral student at the University of Michigan, Castro taught mathematics content and methods courses for elementary teachers.

Hala Ghousseini is a doctoral candidate in Mathematics Education at the University of Michigan-Ann Arbor. She has been working on the BIFOCAL project since its inception, and she also teaches in the teacher preparation program. She earned a bachelor's degree in mathematics and a master's degree in mathematics education from the American University of Beirut in Lebanon. She was a mathematics middle school teacher for 9 years and served as mathematics coordinator for 5 years. Her current research interests focus on the initial preparation and ongoing professional development of teachers of mathematics.

Van Zoest, L. R., Moore, D. L., and Stockero, S. L.
AMTE Monograph 3
The Work of Mathematics Teacher Educators
©2006, pp. 133-148

9

Transition to Teacher Educator: A Collaborative Effort

Laura R. Van Zoest
Diane L. Moore
Western Michigan University

Shari L. Stockero
Michigan Technological University

We use our mentored clinical experience for doctoral students as a context for exploring struggles experienced teachers face as they transition into their new role as preservice teacher educators. Our retrospective analysis identified two consistent issues in our planning sessions, debriefing sessions, and post-experience conversations: preservice teachers' requests for the novice teacher educators to share their experiences as classroom teachers and the novice teacher educators' preference for modeling as an instructional technique. We use dialogue from our sessions to ground our exploration and to illustrate the nature of our collaboration. We conclude by identifying critical components of such experiences.

The vision of teaching mathematics outlined in documents such as the National Council of Teachers of Mathematics *Standards* (1989, 1991, 2000) has required mathematics educators to rethink teacher education. Although there has been much thought about how teacher education should change, little attention has focused on how to prepare mathematics teacher educators who will need to provide this education. In this chapter, we describe one university's attempt to prepare mathematics teacher educators and discuss the struggles three novice teacher educators faced as they transitioned into their new role.

Most of the doctoral students in our program who are interested in becoming teacher educators are experienced school teachers. As such, they bring with them a wealth of knowledge about curriculum, how students think, how to facilitate learning, classroom management, and other instructional practices. They have formed opinions about "what works" in the classroom and often have strong beliefs about how students learn. Much of this knowledge is not based on research, but rather on their own experiences. Although school teaching experience can certainly be an asset, these experiences alone will not ensure success as a teacher educator. The transition from experienced school teacher to teacher educator is complicated and does not occur automatically or quickly. In fact, Murray and Male (2005) found that the evolutionary process of establishing a new professional identity requires two to three years for most people. A large part of this transition involves identifying how former school teachers can "draw on their accumulated professional knowledge and understanding of school teaching to achieve feelings of personal confidence about inducting student teachers into the profession" (Murray & Male, 2005, p. 136).

To ease that transition, our program requires doctoral students in mathematics teacher education to take a three-credit, semester-long course designed to provide an in-depth experience with the thinking, planning, and reflecting involved in teaching a mathematics methods course. Each student is assigned to a methods course at either the elementary or secondary level, depending on interest, and works with the faculty member who is teaching that course. Although the specifics of the experience have varied with individual faculty mentors and the past experiences and needs of the doctoral students, constants include attending each session of the methods course, participating in planning and debriefing with the faculty mentor, and interacting with preservice teachers. Recently, doctoral students have also been required to independently teach the same methods course under the supervision of the faculty mentor in a subsequent semester.

During the Spring 2004 semester, three doctoral students were interested in fulfilling this requirement at the secondary level. That semester Laura was teaching both the middle and high school methods courses in our secondary mathematics teacher education program. The mathematics education faculty decided to have all three work with her in the middle school methods course as its structure provided the greater opportunity for examining preservice teacher thinking. The faculty also felt that it might be advantageous to put the doctoral students in the same class so they could interact with each other as well as with their faculty mentor.

In this chapter we share some of what we have learned from this experience. We also use this experience as a context to identify critical components in the design of the "mentored clinical experiences that develop expertise in designing and teaching mathematics content and methods courses for teachers" recommended in the Association of Mathematics Teacher Educators' *Principles to Guide the Design and Implementation of Doctoral Programs in Mathematics Education* (2002, Part II, ¶8).

The Participants and Context

The Participants

Laura had taught the middle school methods course for three consecutive semesters and welcomed the opportunity to have experienced school teachers join her in rethinking the content of the course. She was confident that the collective expertise of the teachers would greatly enhance what could be offered to the preservice teachers. All three teacher education doctoral students—Karen, Shari, and Diane—were well-respected, competent, accomplished high school mathematics teachers who had each spent time reflecting on her own teaching, working on state standards initiatives, and facilitating sessions to help colleagues think about teaching and learning.

Karen's former position as department chair had required her to mentor all the teachers in her department as they implemented a National Science Foundation-funded reform mathematics program developed by the Core-Plus Mathematics Project (CPMP). Diane had received the Presidential Award for Mathematics and Science Teaching, and had just recently earned National Board Certification in secondary mathematics teaching. As professional developers at the regional and national levels, respectively, Karen and Diane's role had been to support districts that had adopted CPMP by increasing the fidelity with which the teachers implemented the curriculum. They used student materials to engage teachers as learners of the mathematical content and drew on teacher and implementation guide materials, as well as their own experiences as teachers of the curriculum, to address concerns and issues of classroom management, pedagogy, and beliefs about teaching and learning.

After ten years at the high school level, Shari took a position as coordinator of the first year mathematics program at a state university and mentored the predominately international graduate assistants who taught mathematics classes to incoming freshmen. This included teaching a course for the teaching assistants that focused on quickly preparing them to implement a departmentally-defined, fast-paced curriculum in a way that engages undergraduate students. In short, each of the three practitioners brought to the collaboration the important attributes of maturity, commitment, and experience described by Labaree (2003) as strengths for those working toward a doctorate in mathematics education.

None of us anticipated the extent to which our pedagogical skills would be challenged or the degree to which our understanding of teaching and learning would need to be extended. During the semester's collaboration, we (the doctoral students and the faculty mentor) constantly questioned our practices as we struggled with issues of how to facilitate teacher learning and how best to use past experience and knowledge of the teaching profession to support that learning. Despite their extensive experience as mathematics teachers, mentors, and professional developers, the three teacher practitioners were positioned as expert become novice (Murray & Male, 2005) as they began their university doctoral program and their work with preservice teachers. Furthermore, Laura's skills as a teacher educator were challenged as she rethought what it meant to develop as a teacher educator and how to facilitate that in novice teacher educators.

The Expectations

At the beginning of the experience, the following expectations were established for the participating doctoral students:

1. Attend all sessions of the middle school methods course and read all the readings that are required of the undergraduate students enrolled in it.
2. Participate in a weekly 2-hour planning session.
3. Participate in a weekly 2-hour debriefing session.
4. Participate in grading major assignments and in giving ongoing feedback to the undergraduate students.
5. Participate in teaching selected class sessions at the discretion of the instructor.

These expectations were adhered to throughout the course. From the beginning, Laura's interest in utilizing Karen, Diane, and Shari's expertise led to a greater co-construction of both the lessons and their implementation than was necessary to satisfy the course expectations.

The Middle School Methods Course

The middle school methods course is the first of three mathematics methods courses for secondary mathematics education majors and also is taken by all secondary mathematics education minors. The course focuses on teaching for student understanding by accessing and building on student thinking. A key component of the course is a group field experience with an intensive reflection component (see Van Zoest, 2004, for more information). During the Spring 2004 semester, there were fifteen majors and five minors in the course, most in their junior year of college. Class materials included written and videotape cases of teaching, student work, and readings from a range of mathematics education publications.

The Collaboration

Our work together began with Laura providing an orientation to the goals of and parameters for the methods course. Because of her interest in drawing on the doctoral students' experience, Laura shared what she had done in the past as a starting, rather than finishing, point. Our planning sessions included discussions about instructional activities that would best support the preservice teachers' learning and ways to effectively assess that learning. The debriefing sessions mirrored the planning sessions in their emphasis on whether or not the undergraduate students were progressing as we had hoped and whether the activities and assessments we had chosen had been successful in moving them forward on their individual learning trajectories. There were representative undergraduate students who dominated our discussions; however, over the course of the semester, each member of the class was mentioned by name.

The methods course sessions were videotaped, copies were kept of the preservice teachers' major assignments, and the majority of our planning and debriefing sessions were audiotaped and transcribed. This data allowed us to conduct a retrospective analysis of our collaboration. During this analysis, we identified a consistent theme in both our planning and debriefing sessions, as well as our post-experience conversations: the role of classroom experience in teaching preservice teachers. In particular, our analysis revealed that we spent a significant amount of our time struggling with two issues: the preservice teachers' requests for the novice teacher educators to share their experiences as classroom teachers and the novice teacher educators' preference for modeling as an instructional technique. We use dialogue from our sessions to ground our exploration of these issues and to provide a glimpse into the nature of our collaboration.

Tell Me the Answer

In a survey of novice teachers, Smith (2005) found that 72.5% of the teachers surveyed thought that possessing recent classroom experience was important to being a good teacher educator. Adding to this, Murray and Male (2005) found that the majority of the new teacher educators in their study emphasized their classroom teaching experience to their undergraduate students. The novice teacher educators in our collaboration struggled with how sharing their classroom teaching experience fit with our goal of problematizing teaching to develop the preservice teachers' ability to make critical decisions about how to support student learning. This became an issue when the preservice teachers viewed the former teachers' collective experience as a vehicle for finding answers without the hard work of constructing understandings. Karen described it this way:

> Cause Adam was asking me a bunch of questions, like, "So you have the kids in groups every day?" "Yep." And then he was asking me, "Well, then how do you do it when there's not one teacher with each group?" And so I kind of explained some of my strategies for getting around to the groups and things. But he kept asking that and he really wanted to know how that can work. "Do you cover all the topics you should?" And so we got into that a little, with not the breadth, but the depth kind of thing. (2/4/04 Debriefing)

The preservice teachers requested, and expected, answers about how to handle particular teaching and learning situations. Although the novice teacher educators complied with this request on some occasions, as seen in the excerpt above, they tried consciously to refrain from doing so. Some preservice teachers expressed frustration that the novice teacher educators were not sharing their experiences in the way that they wanted. Nevertheless, the novice teacher educators were convinced of the negative impact of complying with the preservice teachers' requests. We see this in Diane's comments at debriefing sessions over a month apart:

> I think it's really powerful also to not set myself out there to know the answers, but just to have more questions. Because I think it's the same idea that Evan was talking about [regarding his relationship with his middle school student group]—we're in this together, and I don't know all the answers, I just know a whole bunch of questions. I have some experience that will help me understand some answers at this point. (2/4/04 Debriefing)

> In fact, I intentionally did not answer questions that Rick wanted a direct answer for. Like he'll say, "What do the three of you think, with your experience?" And I purposely have not responded to those questions because I don't feel like I should give him all the answers. And what are my answers anyway? They're just based on my own experience. (3/10/04 Debriefing)

If preservice teachers held a view of teaching as a set of "tricks," we were concerned that they would be less likely to become reflective practitioners who would carefully evaluate their instructional decisions on an ongoing basis. In some ways, the repeated verbalizing of these reasons seemed to strengthen the novice teacher educators' resolve to have the preservice teachers develop their own abilities to think through solutions to teaching dilemmas. However, the novice teacher educators did not recognize the conflict between this resolve and their desire to show the preservice teachers examples of effective teaching.

Showing How It's Done

As we considered how to prepare the preservice teachers for their second field experience, a subtle tension between Laura and the novice teacher educators surfaced around how to engage the preservice teachers with instances of classroom instruction. Would it be more effective to show a video of a preservice teacher (Ryan) from a previous semester's field experience or to show a video of an experienced teacher teaching a similar lesson to a whole class? Here is a segment of the dialogue from our 2/2/04 debriefing session:

Diane: So here we have an experienced teacher, talking about, launching it, watching the kids at work, reflecting on it herself, and then closing the lesson. That's my dilemma, do you show people who are more experienced and talk about why that worked, or how what they did could be useful to us?

Karen: Maybe, I think at this point the experienced teacher might be helpful to them, where as before we were showing them your [undergraduate] students from last semester so they had an idea of what their experience was going to be like, because they hadn't done it yet.

Diane: Right.

Karen: So now that they have got that experience they know what it's like. Then maybe the more experienced teacher is valuable. To see what can be done.

Shari: Yeah, I don't know if Ryan's tape is going to be as valuable now. Because I think they already know the situation now.

Laura: Yeah, I think the only thing that it would be useful for is if we want to have a concrete instance for everyone, a common experience for us all to react to. You know, we know for example that Ryan is trying to get at their thinking, and to say, "Alright, we couldn't all watch each other, but here is a tape we can watch. Let's watch this five minutes, and then we can talk about what we might have done in this situation, what seemed to work well, how it relates to your experience." So that's using it a different way. That's the way we can use it.

During the 2/4/04 planning session Ryan's videotape was reintroduced to the conversation:

Shari: You know, I was just thinking about [the difficulty of building on undergraduate students' thinking]. What we had from Ryan's class, where they were talking about the area and the one girl counted perimeter.

Karen: Want me to grab [Ryan's videotape]?

Shari: Yeah, I was just thinking of that as an example... And it was a good example, she cleared up her own thinking because he made her say it out loud. He did a good job because he didn't blow it off even though he already had a right answer out there. It was a nice example.

Karen: Yeah, he came back to the incorrect answer. Yeah, they probably still need to see this, I would think.

In the end, we agreed to show Ryan's video, but the novice teacher educators' reasons seemed to be different from Laura's. While Laura saw the video as a focal point for analyzing teaching, the others saw it as a way to model some of the things that we were trying to get the preservice teachers to do in their field experience. After watching Ryan's video and discussing which segment would be the most appropriate, the discussion concluded with Laura saying "I think we could stop it before this, and have a conversation" and the novice teacher educators emphasizing their satisfaction with the fact that the clip was not idealized and served as a realistic example of what our undergraduate students could be expected to do.

In retrospect, this discussion would have been a prime opportunity for Laura to challenge the novice teacher educators to think about their reasons for wanting to provide the preservice teachers with examples of realistic teaching. We know now that all three novice teacher educators had experienced success through modeling specific teaching actions they wanted teachers to replicate. For example, as professional developers focused on implementing specific curricular materials, Diane and Karen had used modeling as a technique to allow teachers to observe the role of the teacher intended by the materials as they experienced the role of the students. Without a specific analysis of the goals and outcomes of these actions, it was difficult for the novice teacher educators to let go of a technique that they had found useful in the past.

Another instance of this tension occurred in response to a concern expressed by some of the preservice teachers that what they had done in the field with a small group of students could not be scaled up to a whole class. We discussed possible ways to address this concern in our 2/25/04 planning session:

Shari: I was wondering if it would be valuable maybe for the first half of class to just model a classroom…have them just be a class for a short time and see what that looks like in a whole group setting, where you have a curriculum that's designed to engage students and to give them the experience of being in a classroom. I mean, obviously, the math won't be challenging to them, have questions—"How would a teacher do it with a whole class?" … Yeah, they're not middle school students, but that's stuff they need—I think some of them need a model though.

Karen: Yeah we do it with adults. Experienced teachers would role play. A lot comes out of it.

Shari: That would be a good idea.

Laura: I'm just trying to think what it would look like, I mean, what it would look like to get the authenticity. Because what we've been doing is having them engage in the mathematics as themselves and pushing them for deeper understanding. Could we do that in context? Can we push them? Do we have enough faith in the materials that even though they're middle school materials, you can push their mathematical understanding through their justifications of what they come up with?

Here we see the novice teacher educators pushing to model a classroom by teaching the preservice teachers as if they were middle school students for the purpose of convincing them that what they had accomplished with a small group of students in their field experience could be done with a whole class. Laura was concerned about the difficulty of maintaining the problematic nature of teaching in this situation and the risk of trivializing both the mathematics and the challenges of teaching it well. One of Laura's challenges was conveying this concern to the novice teacher educators in a way that would cause them to examine their motivations without causing them to reject either Laura or their past experiences.

Lessons Learned

In reflecting on the collaborative experience and reading through transcripts that documented much of what occurred, we have identified three recommendations for others who are embarking on similar experiences: (a) engage in explicit conversations about what it means to be a teacher educator; (b) emphasize experiences that are different from classroom teaching; and (c) make such collaborations a required part of the doctoral program and a recognized component of the faculty mentor's workload. Each of these recommendations is elaborated below.

Teacher Educator Conversations

As a part of our reflection about the experience after the fact, we read articles specifically about becoming a teacher educator (e.g., Borko, 2004; Cochran-Smith, 2003; Labaree, 2003; Murray & Male, 2005; Tzur, 2001; Zeichner, 2005) and engaged in explicit conversations about how one's experience as a teacher is similar to and different from being a teacher educator. If we had taken the time to do this either before or during our collaborative teaching experience, we would have been in a better position to figure out what experience the novice teacher educators were bringing to the collaboration and how to make the best use of it. The readings provide a shared language and a source for confronting novice teacher educators about the basis for their ideas. For example, it was much easier for Laura to challenge the reoccurring emphasis on modeling once we had read Murray and Male (2005) and were aware of novice teacher educators' tendencies to share their experience with preservice teachers directly by showing them how it can be done.

Teacher Educator Experiences

In retrospect, Laura's interest in utilizing Karen, Diane, and Shari's expertise in developing the course seemed counterproductive to their development as teacher educators. Their involvement in the selecting of tasks, structuring of the lessons, and the development of assessments meant that they were spending significant amounts of time engaged in activities that were similar to those they had experienced as school teachers. This limited their ability to change the lens through which they studied teaching and learning and to make the transition from the normative perspective they had developed as school teachers to the required analytical perspective of teacher educators (Labaree, 2003).

We feel that the novice teacher educators' time was spent more productively when they were observing, analyzing, and discussing classroom interactions. When we did this, Laura was able to focus the novice teacher educators on analyzing the preservice teachers' thinking. An example of this occurred when Laura selected segments from a lesson that highlighted facilitation issues for each of the co-facilitators. In the following excerpt, she initiated the discussion of Shari's clip with this statement, "The question that I'd like us to think about now is what exactly is he thinking [during this interaction]? What was Ian thinking about?" After watching the clip, the following discussion ensued:

Laura: Okay, so what is it that he was thinking about? Based on what he said, what seems to be …

Shari: He seems to be thinking that doing it [focusing on student thinking] is all about helping the teacher rather than helping the students.

Laura: And why is he thinking that?

Shari: I think he feels like he's not going to get anywhere. He feels like he's not moving along through the problem. By doing this, we're slowing down the students' progress. We are gaining information at the students' expense.

Laura: So what is it that's making him think that?

Diane: I think he sees teaching different from student learning. He's not teaching in this scenario… "Even if I know how students are thinking, how does that help me teach?" Teaching is disconnected in his mind from student thinking and from learning.

Shari: I don't know if he thinks teaching is disconnected from student learning, I think he thinks that if you're teaching well, that [the students] are going to learn it. So, in his mind I don't think he sees understanding student thinking and teaching as disconnected. I think in his mind, if he teaches it the right way, they are going to learn it. Maybe the disconnection is between learning and understanding.

Laura: Me, the teacher, understanding the student thinking and learning?

Shari: No, the students' learning versus their understanding. I think maybe learning and understanding–I think if the students can do stuff, he thinks they understand it.

Diane: So if they can do it, they've learned it and if they've learned it, they understand it. That kind of thing?

Shari: Yes. Doing is understanding. Multiplying by the reciprocal is understanding.

Laura: Well, if you got through it, you did it.

Shari: Not just get through, but get through and do it right. You have to get the right answers in the process. If you do it and get right answers then you understand it.

Diane: And this idea of looking at student thinking is something that we're doing but doesn't contribute to teaching, in his mind. When he's talking about teaching.

Shari: Right. Because I think in his mind, teaching is just this script that you have. That you plan, okay, I am going to do this, this, this, and this. And regardless of where the students are or what they're thinking about the problem, just go through your script.

The focus on analyzing instances of our interaction with preservice teachers provided a context for the novice teacher educators to recognize differences in what it takes to understand the thinking of preservice teachers and to translate their experience with listening to and building on students' mathematical thinking to listening to, making sense of, and building on preservice teachers' thinking. Sessions such as this one allowed Laura to "orient the teacher educator's reflection on relationships between their attempts to transform teachers' practices and the effects of those attempts" (Tzur, 2001, p. 275). We also see them as instances of engaging in inquiry within a learning community—something that Cochran-Smith (2003) suggests may be critical to teacher educators' development because of its potential to help them both learn new things and unlearn longstanding beliefs, knowledge and practices that may be counterproductive in their new role.

The time spent on planning and debriefing for instructional purposes, while important, limited the time we had to make explicit the distinction between teaching secondary school students mathematics and teaching preservice teachers pedagogy and to develop the knowledge and skills specific to mathematics teacher education. This led to feelings such as those expressed in the following excerpt:

Shari: In hindsight, I can think of a lot of different things we could have done.

Diane: I know. And it seems like we're always getting there, which is good, it's a good learning experience for us. But it seems like it…

Shari: It's always in hindsight, though.

Diane: I know. And it's really not any different than what we're asking them to do. Because they need to anticipate what students are going to think. We need to anticipate how they're going to think about it. And it's hard to anticipate without having the experience. It's really the same situation that we're in.

We hypothesize that spending more time upfront on the transition to teacher educator would have allowed Shari, Diane, and Karen to participate in the collaboration as teacher educators rather than as experienced school teachers with different students.

Time for Collaboration

The time involved in thinking hard about preparing teachers is substantial and would have been prohibitive if the experience had not been part of the graduate students' workload. Even the four hours each week that we had scheduled to meet outside of the methods course was often insufficient to accomplish all our goals. As a consequence, we either had to leave issues unaddressed or extend our meeting time. Making the experience a recognized part of the faculty mentor's workload would allow for the time necessary to think about the novice teacher educators' learning and how to facilitate it. For example, the sessions that were most productive involved Laura watching videotape of the novice teachers' instruction, selecting key instances, and formulating questions to stimulate reflective thought. Doing this in addition to teaching the focus course and meeting with the novice teacher educators to plan and debrief warrants a reduction in other responsibilities. If universities are serious about preparing teacher educators who will be able to prepare the kinds of teachers recommended by the *Standards*, they will need to allocate resources accordingly.

Conclusion

The novice teacher educators were placed with Laura in the middle school methods course because its structure provided an opportunity for examining preservice teacher thinking. However, including the novice teacher educators as co-designers of the course shifted our attention to generating student thinking rather than analyzing it. We now have a better understanding of the complexity of the transition from school teacher to teacher educator and the crucial need to deliberately facilitate that shift. Based on our experiences, we have identified several critical components of a mentored clinical experience for novice teacher educators.

First, reading what is known about becoming a teacher educator and having explicit discussions about what it means to be a teacher educator sets the stage for understanding the complexities of the process. Although we discussed readings only during the retrospective analysis of our work together, we recommend that others use such readings as a foundation for the collaboration process. The readings listed in the *Teacher Educator Conversations* section provide a starting point. Second, breaking the familiar cycle of

planning → instructing → assessing → debriefing

creates space for recognizing differences between teaching mathematics and teaching someone to teach mathematics. This can be done both through watching and analyzing someone else's teaching with them and through submitting one's own instruction to such analysis. Essential requirements seem to be an instance of practice and a mentor able to push for the level of analytical thinking required of teacher educators. Third, maintaining a focus on preservice teacher thinking and having explicit conversations about the intent of preservice teacher instruction keeps the novice teacher educators grounded in their new role and forces reflection on the complexities of inducting new teachers into the field.

We encourage others to learn from our experience as they develop programs for teacher educators at their institutions. Benefits for novice teacher educators, preservice teachers, and mentor teacher educators will occur through programs designed to provide experiences that capitalize on novice teacher educators' experience and use it to catalyze movement along the individual learning trajectories of everyone involved. We all have much to learn about teaching. It is only through improving the education of teacher educators that we will be able to improve teacher education systematically.

References

Association of Mathematics Teacher Educators Task Force. (2002). *Principles to guide the design and implementation of doctoral programs in mathematics education*. San Diego, CA: Author.

Borko, H. (2004). Professional development and teacher learning: Mapping the terrain. *Educational Researcher, 33*(8), 3-15.

Cochran-Smith, M. (2003). Learning and unlearning: The education of teacher educators. *Teaching and Teacher Education, 19*, 5-28.

Labaree, D. F. (2003). The peculiar problems of preparing educational researchers. *Educational Researcher, 32*(4), 13-22.

Murray, J., & Male, T. (2005). Becoming a teacher educator: Evidence from the field. *Teaching and Teacher Education, 21*, 125-142.

National Council of Teachers of Mathematics. (1989). *Curriculum and evaluation standards for school mathematics*. Reston, VA: Author.

National Council of Teachers of Mathematics. (1991). *Professional standards for school mathematics*. Reston, VA: Author.

National Council of Teachers of Mathematics. (2000). *Principles and standards for school mathematics*. Reston, VA: Author.

Smith, K. (2005). Teacher educators' expertise: What do novice teachers and teacher educators say? *Teaching and Teacher Education, 21*(2), 177-192.

Tzur, R. (2001). Becoming a mathematics teacher-educator: Conceptualizing the terrain through self-reflective analysis. *Journal of Mathematics Teacher Education, 4*, 259-283.

Van Zoest, L. R. (2004). Preparing for the future: An early field experience that focuses on student thinking. In T. Watanabe and D. R. Thompson (Eds.), *The work of mathematics teacher educators: Exchanging ideas for effective practice (AMTE Monograph 1)* (pp. 124-140). San Diego, CA: Association of Mathematics Teacher Educators.

Zeichner, K. (2005). Becoming a teacher educator: A personal perspective. *Teaching and Teacher Education, 21*, 117-124.

Laura R. Van Zoest, Professor of Mathematics Education at Western Michigan University, received her Ph.D. from Illinois State University. Through university courses, school-based professional development, and wide-ranging initiatives funded by the National Science Foundation and Michigan Department of Education grants, she has supported inservice and preservice teachers as they implement the recommendations of the NCTM *Standards*. Her research questions revolve around how one becomes an effective teacher of mathematics—more recently, teacher of mathematics teaching—and the role of learning communities in the process. She is editor of the recent NCTM book, *Teachers Engaged in Research: Inquiry into Classroom Practice, Grades 9-12.*

Diane L. Moore, a doctoral student at Western Michigan University and a fellow in the Center for the Study of Mathematics Curriculum, is a nationally board-certified teacher who retired from teaching mathematics at the secondary level in 2004. Her research interests include curriculum development, curriculum implementation, and the role discourse plays in the enactment of curriculum. She is currently working with the curriculum development team revising the Core-Plus Mathematics Project curriculum.

Shari L. Stockero, Assistant Professor of Mathematics Education at Michigan Technological University, completed her Ph.D. in mathematics education at Western Michigan University in 2006. She previously taught mathematics and physics at the high school level. Her dissertation research focused on the use of video cases of teaching in preservice teacher education, specifically the extent to which such materials can be used to develop a reflective stance towards teaching. She plans to continue to focus her research on preservice and inservice teacher education.

Sztajn, P., Ball, D. L., and McMahon, T. A.
AMTE Monograph 3
The Work of Mathematics Teacher Educators
©2006, pp. 149-162

10

Designing Learning Opportunities for Mathematics Teacher Developers[1]

Paola Sztajn
University of Georgia

Deborah Loewenberg Ball
Teresa A. McMahon
University of Michigan

Educating the diverse group of professionals who prepare mathematics teachers is a challenge. Because mathematical knowledge for teaching (MKT) is one of the main arenas for the work of these professionals, we hypothesize that MKT can serve as a catalyst for communication among them, providing an intellectual space in which to learn about the work of teacher developers. We describe the design of a learning opportunity in which MKT was central—the 2004 Summer Institute organized by the Center for Proficiency in Teaching Mathematics. We propose further investigation related to the role of MKT in the preparation of mathematics teacher developers.

Between 1998 and 2008, approximately 2.4 million new teachers will be needed in U.S. public schools (Hussar, 1999). These new hires will replace teachers who retire or leave the profession and will teach in new classrooms created to handle the increasing number of school-aged children. Recruiting such large numbers of teachers will be harder in certain disciplines. According to Darling-Hammond (2000), certain disciplines will have a shortage of teachers while others will not; mathematics is among the disciplines with a shortage. For example, 82.5% of public urban middle schools and 95% of public urban high schools are already in need of mathematics teachers to complete their

[1] This material is based upon work supported by the National Science Foundation under Grant No. 0227586 to the Center for Proficiency in Teaching Mathematics. Any opinions, findings, and conclusions or recommendations expressed in this material are those of the authors and do not necessarily reflect the views of the National Science Foundation or the Center.

teaching staff (Recruiting New Teachers, 2000), and many districts have begun offering financial as well as other rewards to recruit and train mathematics teachers (Levine & Christopher, 1998, p. v).

Thus, an important challenge currently facing mathematics educators across the U.S. is the need to train an increasing number of well-prepared mathematics teachers. This challenge relates both to the quantity and the quality of the current mathematics teaching force. On one hand, we have a shortage of mathematics teachers and must increase the number of mathematics teachers working with children in schools. On the other hand, we have an increasing push to prepare proficient teachers—that is, teachers who have a profound understanding of mathematics and school mathematics curriculum, who know how children learn mathematics, who display a repertoire of instructional routines, and who can design and assess instructional interventions (National Research Council, 2001). Data from the National Assessment of Educational Progress (NAEP), for example, showed that "grade 8 students had a significantly higher average scale score when their teachers held a mathematics certification, [but] in more recent years fewer students have benefited from having teachers who hold such certification" (Grouws, Smith & Sztajn, 2004, p. 227). Although certification is not necessarily synonymous with quality, lacking certified teachers is a problem in the move towards having proficient teachers in every classroom. Therefore, the improvement of mathematics education depends on building instructional capacity (Cohen & Ball, 1999) in mathematics.

This need for large numbers of highly qualified mathematics teachers creates a rising demand for high quality teacher education and professional development in mathematics. This in turn leads to a challenge: The need to develop a "teaching force" of mathematics teacher educators and professional developers. Consider the problem we face: Improving mathematics instruction depends on the development of mathematics teachers, but the development of mathematics teachers depends on building the capacity of those responsible for preparing teachers at the preservice and inservice levels. However, the education of teacher educators or the development of professional developers is just beginning to be conceptualized (Cochran-Smith, 2003; Stein, Smith & Silver, 1999). "Teacher developers" comprise a vast array of professionals—teacher mentors, district leaders, mathematicians, faculty in schools of education, to name a few—many of whom have had no special preparation for the work of teaching teachers.

In this paper, we address the challenge of preparing professionals who can prepare high-quality mathematics teachers. We begin the paper by framing the issue of educating the diverse group of people who work to prepare mathematics teachers. We claim that mathematical knowledge for teaching (MKT), as defined by Ball and her colleagues (Ball & Bass, 2003; Ball, Thames, & Phelps, 2005; Hill, Rowan, & Ball, 2005), is one of the main arenas for teacher developers' work, and therefore their own learning. Furthermore, we hypothesize that mathematical knowledge for teaching (MKT) can serve as a catalyst for communication among these professionals, providing them with a common intellectual space in which to work on mathematics teacher education. We began to examine this hypothesis by designing and conducting a professional development institute for teacher developers in which MKT was a central focus of the work. In this paper, we describe the design of the institute we conducted, provide examples of the ways in which MKT was investigated and studied by the teacher developers who participated, and share some anecdotal accounts of how they responded to MKT. We conclude the paper by raising questions to be investigated related to the role of MKT in the preparation of mathematics teacher developers.

Developing Teacher Developers

There is no single word or phrase to describe mathematics teacher developers. This professional group includes faculty members in a two-year college who teach mathematics as well as teacher leaders or math coaches in school districts who are engaged in professional development activities in mathematics. University mathematicians do not ordinarily think of themselves as "teacher educators"—yet they often are. University faculty members who teach mathematics methods courses for prospective teachers do not ordinarily think of themselves as "professional developers"—yet they are. Both in early career mathematics content and mathematics methods courses as well as later professional development, teachers explicitly learn the disciplinary knowledge that forms the basis of the content they will teach to students. They also develop ideas about how that content can be taught. This has implications for what mathematics teacher developers need to know or learn.

Few members of this diverse group have had formal opportunities to learn to work with teachers. Trained as mathematicians or as teachers themselves, most teacher developers lack knowledge about teachers as learners. Further, although they may have taught mathematics that is useful for students, they may not have developed the skills to teach mathematics that is useful for teaching. Van Zoest (2005), for example, reports how difficult it can be for accomplished school mathematics teachers to teach prospective teachers in mathematics methods courses; their experience teaching students does not necessarily equip them with the explicit knowledge or skills to teach teachers. Like teachers, teacher developers of all backgrounds need professional formation. Without deliberate opportunities to develop knowledge and skills, they are not fully prepared to work with teachers.

Although an emerging literature base on the learning of teachers of mathematics exists, there is essentially no research literature on the learning of mathematics teacher educators and professional developers. And, although research on teacher learning is a useful starting point for thinking about the development of those who teach prospective or practicing teachers, one cannot draw direct parallels. Among the questions to be answered are: What do teacher developers know and believe? What do they need to learn? What is challenging about their work that many do not learn simply from experience? What content knowledge do they need? What do they need to understand about the work of classroom teaching? What do they know about their own learners and about how to relate to them effectively? How do they understand the challenges of attending to equity in their teaching?

In order to answer these questions, we argue that one must return to the work of teaching mathematics, for it is toward this work that teacher developers must orient their work with teachers. Successful opportunities for teachers' learning are likely to be ones that increase their capacity for the work that teachers do with students; teacher developers' sense of that work can help inform and ground their efforts with teachers in their courses, workshops, and programs. One major domain in which teachers must learn for their practice, and on which much professional development focuses, is teachers' mathematical knowledge. Our hypothesis is that this domain is a fruitful one in which to work with teacher developers, both because it is likely to be central to their work, and also because we think it can create a medium for collective work that engages and takes advantage of the diversity of this professional group.

Mathematical Knowledge for Teaching

Our concept of "mathematical knowledge for teaching" draws on the work of Ball and her colleagues (Ball & Bass, 2003; Ball, Thames, and Phelps, 2005; Hill, Rowan, & Ball, 2005). It represents a view of teacher content knowledge that is rooted in practice—in the tasks that teachers do that require mathematical knowledge, reasoning, and insight, and in the ways in which teachers have to use mathematics to do those tasks. In this view, teaching mathematics is mathematically demanding work, and the uses to which teachers must put mathematics require them to know a different mathematics, and to know it differently than do other professionals who use the subject. Teachers need to know and understand mathematics in ways directly related to the work of teaching—for example, designing good tasks, diagnosing the difficulties students are having, and managing a productive discussion of mathematics in class. They need to know the "insides" of mathematical ideas as well as the connections among those ideas. They also need the skill to grapple successfully with mathematical issues they might encounter during a lesson. Examples include deciding which mathematical representation might be most fruitful at a particular point in students' learning, determining whether a student's unconventional idea is mathematically plausible or reflects a misconception, and choosing numbers strategically for an illustrative problem.

Ball and her colleagues include as part of the mathematical knowledge needed for teaching the skills required to calculate the following:

$$34 + 148 =$$

They argue that being able to add these numbers and obtain the answer, 182, is something teachers must be able to do. But, they claim, teachers must do more than that. Students might produce answers other than 182, such as:

$$34 + 148 = 172 \qquad 34 + 148 = 488$$

The kind of mathematical analysis involved in figuring out what might lead to these answers involves a kind of reasoning special to teaching, one not needed by other mathematically-educated adults (Ball, Thames, & Phelps, 2005). Similarly, deciding how to represent for students a conventional procedure for adding multi-digit numbers depends on a kind of explicit understanding of the essential place value concepts underlying the algorithm as well as how diverse diagrams or materials might differentially highlight aspects of those concepts. Deciding what numbers to choose to give students opportunities to practice adding numbers, and whether to write the problems vertically or horizontally (as above), or to vary the notation also requires a mathematical perspective and thinking particular to teaching. The construct of mathematical knowledge for teaching encompasses the mathematical knowledge, skill, and insight needed for the work of teaching, and it is this construct that was central to the institute we conducted with a group of teacher developers.

MKT and Learning Opportunities for Teacher Developers

The Center for Proficiency in Teaching Mathematics (CPTM), a NSF-funded alliance of the University of Georgia and the University of Michigan, focuses on ways to support the learning and practice of mathematics teacher developers. The Center designs and studies various approaches to support the learning of those who work with mathematics teachers, including collaborative work, study groups, apprenticeships, on-going professional development sessions, and regional working conferences. These models differ in structure, curriculum, goals, materials, and complexity.

During the 2003-04 academic year, CPTM researchers, faculty, and graduate students from both mathematics education and mathematics departments worked together to design the format and content of a summer institute for professionals who work with K-8 prospective teachers of mathematics. The eight-day institute, "Developing Teachers' Mathematical Knowledge for Teaching," focused on two key questions: 1) What mathematical knowledge and practices play a central role in the everyday work of teaching? 2) What are promising approaches for helping teachers learn mathematics for teaching and learn to use it in their work?

Some would argue that summer institutes are weak interventions and offer inadequate opportunities for sustained and significant professional learning. However, structure alone does not wholly determine the nature of a learning opportunity for teachers, and we do not assume that a summer institute, in the various forms it might take, is automatically not of value. Moreover, teacher developers are a different group of professional learners, and our work focuses on investigating the formats, content, curricula and tasks that might support the learning of this particular group of professionals.

Advertised through national organizational email lists, the summer program attracted over 120 applicants. The original number of available spaces (40) was expanded to accommodate 65 participants. The diversity of the professionals who work in teacher development was confirmed as we reviewed applications of school, district-level, and private mathematics professional developers as well as instructors and researchers from mathematics and mathematics education departments in two-year, four-year, and research institutions. We were surprised, however, to see applications from a textbook writer and from mathematics specialists from a state department of education. Applications were reviewed with an eye toward ability to describe current work and how what was learned from the experience might be used. Teams were given extra consideration. And, because a goal for our research was to understand how different players attend to learning opportunities, we considered factors of diversity the participants would contribute to the group composition in terms of institution, stage of career, geographic location, degree of isolation in work, experience with teachers, and experience in K-12 classrooms. Taking these criteria into account, 45% of the selected SI 2004 participants were from mathematics departments, while the other 65% came from mathematics education departments, districts, or other agencies. From the participants who work in mathematics departments, 64% taught mathematics content courses only and 36% taught some combination of content and education courses.

Design Goals and Assumptions

To begin testing our hypothesis that MKT can engage and take advantage of the diversity of teacher developers, we designed opportunities for the institute's participants to learn about MKT and explore how to use this type of knowledge in the preparation of future teachers. We were explicit about our belief that MKT is an important knowledge for mathematics teacher developers to acquire. We also wanted participants to think about who prospective teachers are, what they know, what they can do, and what they are capable of learning. Furthermore, we wanted participants to consider ways to plan, organize, and implement instruction in mathematics content courses designed for prospective elementary teachers.

Finally, it was our goal to conceptualize with the group the notion of professional development for teachers of teachers—working together to understand better what mathematics teacher education professionals need to continue to learn and grow. Acknowledging that mathematics teacher educators lack a clear professional identity, part of the goal was to enable this group to begin collaboratively to define who they are, to study how they work, and to explore what kinds of initiatives will help them work on their own professional growth.

The design of the institute was guided further by what is known about teachers and their learning, and the assumption that there would be some parallel since teacher developers are teachers. So, planners had in mind that teacher developers need to: (a) learn in and from practice; (b) share with each other and from their professional experiences; (c) participate in some aspects of the design of their professional development experiences; (d) choose professional development opportunities to work on that are most meaningful to them; and (e) be treated as professionals. Furthermore, we had in mind that one cannot just bring a group of people together (as professional as they might be) and expect learning to happen. Learning opportunities need to be carefully designed and planned for participants at all levels. Support and guidance are needed to promote learning.

Purposeful Design

We assumed that the work and learning of this diverse group of teacher developers would be most productive if it were anchored in a shared experience that provided a common frame of reference for discussions and debate. So, at the core of the CPTM 2004 Summer Institute design was what we call a "laboratory class," which enrolled 20 preservice elementary teachers who were studying mathematics content for teaching. The daily two-hour classes can be compared to a shared specimen for observation and manipulation, and the participants to a research team developing hypotheses and looking for evidence to support or refute claims and assumptions.

Each morning during the institute, the participants observed the lab class, taking notes of evidence of the work the preservice teachers were doing, the work the teacher was doing, and the mathematics being studied. The structure of the institute was designed to support the use of the laboratory experience. Prior to observing a class, participants met to review and comment on the detailed lesson plan for the day, look at the mathematical tasks that would be posed in the class, and identify a focus for their observation of that lesson. Participants were asked to focus their observation on the preservice teachers, the teaching, or the mathematics. Specific questions were asked to scaffold meaningful observations of the laboratory class.

After each class, teacher developers worked in small groups and engaged in a variety of analytic activities designed to support their use of the laboratory class as a site for learning. For example, when participants focused their observation on the preservice teachers, they were given the opportunity to examine prospective teachers' written work. They discussed what the preservice teachers in the laboratory class were attending to and raised hypotheses about the preservice teachers' developing mathematical understanding. An extended lunch offered time for exchange and discussion, which was followed by whole group work on the laboratory class, including initial preparation for the following class period. Participants had the opportunity to make suggestions to be included in the next day's lesson, which they discussed the following morning before the laboratory class.

After the whole group discussion, the teacher developers engaged in self-selected activities. There were options available for small group work, such as to examine K-6 curriculum or resources useful for teacher development, watch videotapes of classroom segments and discuss their effectiveness in teacher education, or study children's mathematical work and its use with prospective teachers. Evenings were used by participants to examine student class journals, to work with the class instructors to design the next day's tasks, or to work in teams on how they might incorporate what they were learning when they returned home.

Across all these activities, the institute design intended to make practice visible and provide opportunities to consider what mathematical knowledge and practices play a central role in the everyday work of teaching. We also wanted teacher developers to consider what approaches help teachers learn mathematics for teaching and learn to use MKT in their own work.

MKT and Teacher Developers: An Example from the Summer Institute

During the institute, opportunities were designed and emerged for teacher developers to engage in discussion about MKT. For the purpose of providing an example, we share episodes related to the first laboratory class that participants attended and the ongoing discussions that followed from that class. The goal of this example is not to be exhaustive but illustrative.

Prior to observing the first laboratory class, participants engaged in a series of tasks and discussions about fractions. Specifically, they were asked to analyze some potential representations for $\frac{3}{4}$, and to view and discuss a video in which pupils in a third-grade class considered and made arguments about whether $\frac{4}{4}$ is greater than or equal to $\frac{4}{8}$. The following day, prospective teachers in the laboratory class engaged in the same activities. During the post-class discussions that day, participants talked about what prospective teachers had done with the activities and their use of mathematical language. Many participants voiced that they had underestimated what prospective teachers would be able to do with the given prompts.

Towards the end of the discussion, one participant raised the question of whether a definition of fraction should have been introduced during the laboratory class and whether the introduction of such a definition would change the effectiveness of prospective teachers' experiences in discussing different representations for fractions. Because the comment sparked such animated debate among the teacher developers and because definitions are an important component of MKT, we decided to continue this line of discussion. The next day, during the large group discussion after the laboratory class, participants were asked to write a definition of fractions that they thought would be appropriate to offer the prospective teachers in the laboratory class. We established the criteria that definitions should be mathematically precise as well as usable by the community, that is, based on terms already known and defined. For about an hour, teacher developers from all backgrounds engaged vigorously in small group, and later whole group, discussions about definitions of fractions.

In one small group, a participant offered the definition that "a fraction is a number that can be written in the form $\frac{a}{b}$ where b does not equal zero and both a and b are integers." This definition raised questions of whether fractions are a number or a representation, whether fractions could include variables, and whether it would be best to consider a and b to be whole numbers. In another group, there was a lengthy discussion about whether irrational numbers, such as $\sqrt{2}$ or $\sqrt{5}$, can be used in the numerator and the denominator of fractions. Yet another group discussed whether and how the definition of fraction as a part of a whole related to $\frac{3}{4}$ and to 0.75. Another group of participants engaged in discussions about the relations between ratios and fractions and how the definition of fraction could account for that relation. Woven through all these discussions was the question of what teachers need to know and for what purpose. Similar arguments continued in the whole group discussion that followed, with no final consensus emerging among participants about either what definition of fraction should be used in the context of the class or what such a definition would add to the class.

Throughout the week the discussion of what would constitute a good definition of fraction for the laboratory class—and when and if such a definition should be offered to prospective teachers—continued to constitute an issue for informal discussion among participants during lunch hours and in the evenings. The discussion of definitions was also included in the laboratory class as the prospective teachers were asked to engage in a discussion about what would constitute good definitions for mathematics instruction. The conversation continued in emails after the institute. Some participants were critical of the laboratory class instructor's decision not to explicitly introduce a definition of fraction. One participant argued that prospective teachers were ripe "to start incorporating the mental discipline required by the successful introduction of the definition of fractions" and were denied such opportunity. Another participant, a research mathematician, commented he had never spent so much time thinking about fractions and had never thought he would need to do so. Other participants wrote about thinking about fractions as numbers versus fractions as a notation to denote an operation.

As they work with teachers, teacher educators continually will face questions such as what constitutes a useful definition of "fraction" as well as why and how to make tasks that provoke understanding of the meaning of fractions. We did not, however, use this example to offer a definitive answer to whether or not definitions of fractions should be used with preservice teachers—even if we wanted to offer a final word, such a definite answer is not available. Rather, the reason we share this example is to illustrate the ways in which MKT engaged a large group of mathematics teacher developers, with many different backgrounds and a variety of perspectives, in meaningful and ongoing discussions. We believe this example, and others that could be extracted from the Summer Institute experience, begin to show how MKT can spark meaningful conversations among the diverse members of this professional community about issues of content that are at the heart of their work as mathematics teacher developers. MKT is a productive focus for their own learning as well as for that of the teachers with whom they work.

Conclusion

We started this paper considering the need to develop a teaching force of mathematics teacher developers and the fact that most of those who work in such capacity have no formal training in educating teachers. We suggested that returning to the work of teaching mathematics is important and discussed the concept of mathematical knowledge for teaching, presenting this concept not only as an important knowledge for teacher developers but also as a concept that can help the diverse professional group of teacher developers come together. We hypothesized that MKT could serve as a catalyst for communication among teacher developers from diverse backgrounds and professional experience. To begin exploring this hypothesis, we presented the goals and the design of a summer institute for teacher developers in which MKT was a central component. We shared anecdotal experiences from the institute and illustrated how ideas about the mathematical knowledge teachers need for their work can spark meaningful professional conversations among mathematics teacher developers.

Many questions still remain as we continue to explore our hypothesis about the role of MKT in educating teacher developers. For example:

- How did different features of the design of the institute facilitate the discussion of ideas about MKT among teacher developers?
- How did teacher developers of different backgrounds engage in the activities of the summer institute—observing the laboratory class, discussing the prospective teachers' work and the mathematical tasks in which they engaged, working on the mathematical ideas themselves—and what did they seem to make of what they did?
- Were there mathematical ideas that were particularly challenging for teacher developers to think about in the context of considering mathematical knowledge for *teaching* (MKT)?
- What MKT concepts lead to consensus among mathematics teacher developers and what concepts lead to divergence? What is the role of consensus and dissent in generating conversations and a sense of professional community among mathematics teacher developers?

To continue exploring these questions, we are engaged currently in a systematic analysis of the CPTM 2004 Summer Institute data. We documented the program closely, using multiple tools, media, and sources. The data archive includes participants' individual work, high-quality videotape of most sessions, observations of group and individual activities, and participants' notes and writings.

The conversation about educating teacher developers is just beginning. However, its importance cannot be minimized, for the improvement of mathematics instruction depends on the preparation of proficient teachers. As other venues and formats for educating teacher developers continue to emerge, it is important to keep in mind the development of the professional identity of this group. As teacher educators and mathematicians engage collaboratively in formulating, testing, reflecting on, and modifying ideas for their own professional development, they will simultaneously extend their capacity to use those ideas productively with prospective and practicing teachers.

References

Ball, D. L., & Bass, H. (2003). Toward a practice-based theory of mathematical knowledge for teaching. In B. Davis & E. Simmt (Eds.), *Proceedings of the 2002 Annual Meeting of the Canadian Mathematics Education Study Group*, (pp. 3-14). Edmonton, AB: CMESG/GCEDM.

Ball, D. L., Thames, M., & Phelps, G. (2005). *Content knowledge for teaching: What makes it so special?* Paper presented at the annual meeting of the American Educational Research Association, Toronto, Canada.

Cochran-Smith, M. (2003). Learning and unlearning: The education of teacher educators. *Teaching and Teacher Education, 19* (1), 5-28.

Cohen, D., & Ball, D. L. (1999). *Instruction, capacity, and improvement* (Research Report Series No. RR-43). University of Pennsylvania: Consortium for Policy Research in Education.

Darling-Hammond, L. (2000). *Solving the dilemmas of teacher supply, demand and standards.* National Commission on Teaching and America's Future. Available: http://www.nctaf.org/article/index.php?c=4&sc=17&ssc=0&a=21 & [2005, June 17].

Hill, H., Rowan, B., & Ball, D. L. (2005). Effects of teachers' mathematical knowledge for teaching on student achievement. *American Educational Research Journal, 42*(2), 371-406.

Hussar, W. J. (1999). *Predicting the need for newly hired teachers in the United States to 2008-09.* NCES, document 1999026. Available: http://nces.ed.gov/pubsearch/pubsinfo.asp?pubid=1999026 [2005, June 17].

Grouws, D., Smith, M. and Sztajn, P. (2004). NAEP findings on the preparation and practices of mathematics teachers. In P. Kloosterman and F. K. Lester, Jr. (Eds.), *Results and interpretations of the 1990 to 2000 mathematics assessments of the National Assessment of Educational Progress* (pp. 221-267). Reston, VA: National Council of Teachers of Mathematics.

Levine, R., & Christopher, B. (1998). *Public school districts in the United States: A statistical profile, 1987-1988 to 1993-1994.* NCES, Statistical Analysis Report.

National Research Council. (2001). *Adding it up: Helping children learn mathematics.* J. Kilpatrick, J. Swafford, B. Findell (Eds). Mathematics Learning Study Committee, Center for Education, Division of Behavioral and Social Sciences and Education. Washington, DC: National Academy Press.

Recruiting New Teachers, Inc., Urban Teacher Collaborative, and The Council of the Great City Colleges of Education (2000). *The urban teacher challenge: Teacher demand and supply in the Great City Schools.* Report, Recruiting New Teachers, Inc. Belmont, MA.

Stein, M. K., Smith, M. S., & Silver, E. A. (1999). The development of professional developers: Learning to assist teachers in new settings in new ways. *Harvard Educational Review, 69* (3), 237-269.

Van Zoest, L. R. (2005, April). *Intentional teacher education preparation.* Symposium conducted at the Research Presession of the annual meeting of the National Council of Teachers of Mathematics, Anaheim, CA.

Paola Sztajn is an Associate Professor of Mathematics Education at the University of Georgia and a researcher within the Center for Proficiency in Teaching Mathematics. She works with prospective and practicing elementary mathematics teachers to improve the mathematics instruction of all children. The overarching questions that guide Sztajn's research agenda are: what knowledge do elementary mathematics teachers need to teach high quality mathematics to all children? In which ways do practicing elementary mathematics teachers acquire and continue to develop this knowledge? She is interested in collaborative studies that allow in-depth investigations of such complex questions.

Deborah Loewenberg Ball is Dean of the School of Education and William H. Payne Collegiate Professor at the University of Michigan. Ball's research draws on her many years of experience as an elementary classroom teacher. She studies mathematics instruction and the content knowledge needed for effective instruction. Ball also directs several research projects that investigate efforts to improve teaching through policy, reform initiatives, and teacher education – including the Center for Proficiency in Teaching Mathematics (CPTM), which she co-directs with Jeremy Kilpatrick and Patricia Wilson of the University of Georgia.

Teresa A. McMahon is an Assistant Professor at the University of Michigan and a researcher within the Center for Proficiency in Teaching Mathematics (CPTM). McMahon studies how prospective and practicing teacher educators improve their instruction. Of particular interest is how to support professional community and collaboration so that each might contribute significantly to the development process.